Sustainable Buildings from Marble Waste Heaps

Katia Talento

Sustainable Buildings from Marble Waste Heaps

A New Methodology for Landscape Regeneration and Reuse

Springer

Katia Talento
Instituto Superior Técnico, CERIS Research Centre
University of Lisbon
Lisbon, Portugal

ISBN 978-3-031-72145-8 ISBN 978-3-031-72146-5 (eBook)
https://doi.org/10.1007/978-3-031-72146-5

This Springer imprint is published by the registered company Springer Nature Switzerland AG
The registered company address is: Gewerbestrasse 11, 6330 Cham, Switzerland

Marble quarry and associated waste-heap in Vila Viçosa Municipality (Portugal)

To the difficulties that open to new scenarios
and opportunities.
To all young architects, don't give up.
To dreams, the driver of all.

Acknowledgements

I would like to acknowledge the **Fundação para a Ciência e a Tecnologia** (FCT) who assisted my research financially, under Grant numbers BD/133777/2017 and COVID/BD/152171/2021.

Equally important has been the support of **Professor Miguel Amado**, who guided and supported me in all the decisions taken during my research work, and **Professor José Carlos Kullberg**, who helped me with the geological aspects of the research, the work on site, the detailed survey of a selected quarry in the Estremoz Anticline and the writing of Sects. 5.1 and 5.4.2 of this book.

I am grateful for the inspiring discussions with **Professor Biagio Cillo**, who gave me invaluable advice and suggestions of literature.

I am also extremely thankful for the stimulating conversation with **Doctor Vitor da Silva**, who helped me broaden my horizons about planning and landscape rehabilitation concepts, guiding me through the process of finding the most innovative approach to go ahead the research and motivating me to do more and better.

My great thanks go to **Assimagra Association** that, through the sharing of its work, helped me to identify the quarries (belonging to the study area) on territorial maps and to physically deal with them in ArcGIS™ Software.

Thank you to **João** for having lit hope in a moment when everything seemed dark, thank you to all those who directly or indirectly, in some way, contributed to the realization of this work, and thank you to my **family**, last but not least, for always being by my side even 2700 km away.

About the Book

The Estremoz Anticline, in Portugal, is one of world's main sites of marble extraction. Annually, millions-of-tons of marble are extracted from the multiple quarries present in the area but only about 9% of picked material is marketed because only this percentage is considered of "quality". Due to economic and geological reasons, today, only a small number of quarries remains active, and this situation has contributed to the degradation of the area.

The remaining 91% of the unused material is shelved in stone accumulations, designated by "*escombreiras*", creating a strong impact in the landscape, since they cannot be disposed elsewhere due to associated costs.

The current research challenges the traditional view of the waste heaps as mere byproducts with no value, proposing instead to preserve the historical and industrial memory of the site by integrating the waste into the landscape rather than removing it. The study redefines these accumulations not as imperfections but as elements with significant eco-friendly potential, which can play a crucial role in the area's requalification.

The book delves into the broader implications of human activity on the environment, particularly focusing on the extraction industry's role in landscape transformation. It highlights the urgent need to rehabilitate degraded industrial sites, exploring various strategies for repurposing them into areas of architectural and environmental value. It also presents several case studies of quarries that have been successfully transformed into public parks, cultural spaces, and tourism facilities, underscoring the economic and environmental benefits of such interventions.

Furthermore, the book examines the issue of massive waste accumulations resulting from extraction activities, which have traditionally been excluded from rehabilitation processes. The research advocates for a shift towards an integrated approach, where waste materials are considered an integral part of the regeneration process of an extractive spot. By developing a practical reuse methodology, the book provides innovative solutions for quarry rehabilitation that align with the principles of a circular economy, offering substantial benefits for landscape transformation and sustainable land use.

Through a detailed discussion of landscape theory, quarry and waste reuse, and circular economy concepts, this research provides a comprehensive survey of practical examples and methodologies for requalifying industrial sites, including the reuse of waste heaps. This culminates in the creation of a scientific formula and guidelines for planners, which can be applied to the Estremoz Anticline and similar sites worldwide, filling an existing gap both in theoretical and practical sectors. The outcomes demonstrate the potential for transforming waste heaps into functional and valuable spaces, thereby promoting a sustainable and livable environment while setting a solid foundation for future investigations in the field.

Contents

List of Figures

List of Tables

Acronyms

UN	United Nations
SDGs	Sustainable Development Goals
SLOAPs	Spaces Left Over After Planning
UNESCO	United Nations Educational, Scientific and Cultural Organization
LOD	Wastelandscape of Dwelling
LOT	Wastelandscape of Transition
LIN	Wastelandscape of Infrastructure
LOO	Wastelandscape of Obsolescence
LEX	Wastelandscape of Exchange
LOCO	Wastelandscape of Contamination
LMBV	Lausitzer und Mitteldeutsche Bergbau-Verwaltungsgesellschaft
TNT	Trinitrotoluene
EPA	United States Environmental Protection Agency
OMZ	Ossa-Morena Zone
LNEG	Laboratório Nacional de Energia e Geologia
PDM	Plano Diretor Municipal
DGEG	Direção Geral de Energia e Geologia
GCPs	Ground Control Points
GSP	Ground Station Pro

Part I

Chapter 1
Introduction

There must be a better way to make the things we want, a way that doesn't spoil the sky, or the rain or the land.

Paul McCartney

1.1 Research Overview

Research proposal foreword, focusing on how the idea of the study eventuated, and describing the chapters that follow

In the middle of the twentieth century, one saw planet earth from space for the first time. From space, it takes the form of a small and fragile sphere dominated by a set of oceans, clouds, vegetation, and soils and not by buildings and human activity. The inability of humans to fit their activities into that set is drastically changing the systems of the planet, and many changes are flanked by life-threatening hazards. This new reality, from which there is no escape, must be recognized and handled. Nevertheless, one hopes for a better future, but this hope is strongly conditioned by a significant political action to ensure sustainable human progress and survival.

More than 11 million hectares of forests are destroyed each year, and this amount, over three decades, would correspond to an area with the size of India. Most of this forest is normally converted to low-quality farmland unable to support the farmers who work on it. The acid precipitations kill forests and lakes, and they damage the cultural and the architectural heritage of nations. The burning of fossil fuels implies the introduction of carbon dioxide into the atmosphere, which is causing a gradual global warming. This "greenhouse effect," in the next century, could increase the global temperature enough to alter agricultural production areas and raise sea levels, causing the flooding of the coastal cities, and the disintegration of the national economies. These are the main effects provoked by humans who struggle for survival and prosperity with little consideration of their impact on other people. Some people, excessively consume the Earth's resources without thinking about what to leave to future generations, while others consume far too little, living with the prospect of hunger, disease, and early death (World Commission on Environment and & Development, 1987).

K. Talento, *Sustainable Buildings from Marble Waste Heaps*,
https://doi.org/10.1007/978-3-031-72146-5_1

From the mid-1950s, in some places of the world, development and growth have significantly increased the living standards and the quality of life of people. Many of the technologies and products involved in this improvement are energy-intensive raw materials that provoke a large amount of pollution. The resulting impact on the environment is bigger than ever before in human history. In the last century, the use of fossil fuels has increased thirty times, and industrial production has increased more than fifty times.

The peak of this growth occurred since 1950, and its impact can be noticed in the supply of the world's consumption of a variety of resource-intensive products. The complexity of human needs for natural resources has significantly increased with the growing levels of population and production. Starting from this premise, the already mentioned "greenhouse effect" comes directly from the uncontrolled and exaggerated use of resources. In fact, the combustion of fossil fuels and the cutting and burning of forests emanate carbon dioxide (CO_2) in the atmosphere, and its accumulation, jointly with other gases, catches the solar radiation near the Earth's surface, producing global warming (WMO, 1985).

For all these reasons, it is important to adopt sustainable development, which "*is the development that satisfies the needs of the present without compromising the capacity of future generations to satisfy their own needs*" (World Commission on Environment and & Development, 1987). Therefore, the satisfaction of human needs and aspirations is the main objective of this development, and beyond the basic needs of people, such as shelter, food, clothing, and jobs, there are reasonable aspirations to improve their quality of life. Thus, the sustainable development involves the promotion of values that stimulate consumption patterns that are within the limits of the possible ecology and to which all people can reasonably aspire, and the reduction of non-renewable energies. For this effect, the world must rapidly create strategies that will allow nations to move from their destructive present towards a sustainable process of growth and development, which will involve policy changes in all countries.

In this regard, in September 2015, the United Nations (UN), brought together world leaders to adopt an ambitious Agenda for the eradication of poverty and to promote economic, social, and environmental development on a global scale until 2030. This Agenda is well known as the "Agenda 2030 for sustainable development," and it represents the result of a joint work of governments and citizens around the world, to create a new global model for the promotion of prosperity and well-being for all, protecting the environment and combating climate change. For these purposes, the UN Agenda 2030 defined 17 sustainable development goals (SDGs), successors to the 8 millennium development goals, which should be implemented by all countries (ONU, 2016).

Although these objectives address the economic, social, and environmental dimensions, they also affect the architecture field, with a particular focus on sustainable architecture. It means that a sustainable architecture primarily is that who compensates for the damage caused to nature and the social structure and the resources spent for the realization of the building. When applying this complex concept to the practical part of the architecture, it then refers to a type of design that creates healthy living environments while aiming to minimize the negative impacts, the use of human resources, and energy consumption. Sustainable architecture is reflected in a building's materials, construction methods, resource use, and design in general (Barker associates, n.d.).

While it must be functional and aesthetically superior, the space must be constructed with the mindset of achieving long-term energy and resource efficiency, improving the use of renewable energies and natural materials, and the reduction of wastes, through the reuse and recycle of what already exists, instead of using new materials. Wastes have always been recognized as unwanted or undesired materials, but they could represent great and innovative elements to design architectures and public spaces, also redefining the meaning of the landscape. The philosopher Malcolm Budd considered the real landscape as the nature not altered by man (Budd, 2002).

Recognizing that the contemporary era has become the symbol of a strong industrialization and the insistent brand of man on the territory, nowadays, is it possible to consider the definition of "untouched nature"?

This conception does not fit in with the current age, where, on the contrary, one has assisted, with a nostalgic feeling, to the loss of the landscape. Indeed, the original territorial structure has been permanently changed by the growing industrialization and its subsequent environmental problems (Veríssimo Serrão, 2011).

Being the industrial activity that causes more impact and metamorphosis on the landscape, and since they have modified, and still modify, the landscape in a more invasive way, the research pays attention to the extraction sites. The effects of the strong industrialization are much more visible in this case, and the result of this situation is not only the inclusion of the industrial buildings on the territory but the entire transfiguration of the morphology and physiognomy of the landscape along with its economy, social structure, and environment quality (Cipriani, 2012) (Talento et al., 2018).

Returning to the principles of sustainable development, previously described, the present research intends to provide a fundamental contribution to promote new architectural strategies in terms of sustainability, focusing on the reuse operation of existing materials. Indeed, the research describes a case study in Portugal, the Estremoz Anticline, where resides the highest concentration of marble quarries and wastes, accumulated in huge piles of material, named in Portuguese as "*escombreiras*." The purpose of the work is to give importance to the concept of waste reuse in landscape and architecture and to preserve the identity of the place. New solutions to rehabilitate the quarries and reuse the mounds of waste, rather than to break down them, are possible. In addition, the research shows the interest in transforming an industrial site into a liveable touristic, cultural, recreational, productive, and naturalistic place, having also aesthetic quality, through the rethinking of the entire area and the reusing of the mounds of wastes, integrating them in the surroundings.

The work presented in this book is divided into three parts. The **first part** represents the introduction of the work, and it defines the background, which gives an insight into the topic of the quarries, describing their main features and intrinsic problems. It also discusses the research sphere, scope, problematics, objectives, questions, hypothesis, and methodology. The **second part** includes the knowledge acquired through the analysis of academic and scientific documents about the topic of landscape mutation and the rehabilitation of quarries. Finally, the **third part** addresses the question of waste reuse and circular economy, elaborating, from an innovative methodology of reuse, based on scientific criteria, a formula of practical use for the rehabilitation of quarries and the reuse of waste heaps. This part also presents the case study, and it proposes new sustainable strategies for the replanning of the overall area under investigation and the design of an architectural proposal, in accordance with the derived reuse principles.

1.1.1 Background

From the definition of minerals to the description of the ongoing problems resulting from the extraction activity

The current research, as previously described, addresses the issue of the rehabilitation of a system of marble quarries. Specific clarifications are the keys to the full understanding of this concept. In this regard, the present section aims to provide some important definitions: what is a quarry, what is a mineral, how many categories of minerals exist on earth, which are the different types of quarries based on different cultivation methods, what are the aspects that condition the development of the cultivation project, what are the effects of the extraction industry in the landscape and, after its exploitation, in what turns a quarry, and what is its current situation.

1.1.1.1 The Quarry Industry: Definitions

As previously mentioned, and belonging to the scope of this research, the work focuses on the analysis and the description of the quarry, which is any site of open-air excavation of rocks and minerals (Boca & Oneto, 1989; Brodkom, 2001; Fornaro et al., 2002; Frare, 1996; Vallario, 1998). Apart from the more artisan aspects, a quarry can look like an organized industrial complex to economically exploit a deposit, on the surface or at a small depth, by means of works that take place at the surface of the ground (Gerbella, 1960). “Cultivating” comes from the Latin “*colère*” (p.p. *cultum*), which has various meanings, such as cultivating the land, cultivating friendships, cultivating the arts and studies, respecting a person, and honouring the habits of people. The excavation of mineral resources is an activity, like many other crops (agricultural, forestry, fishing, water, etc.), through which raw materials are extracted from natural resources; these subjects have been necessary for Man since the dawn of civilization (Gisotti, 2008). Minerals are mainly relevant in every sector, from that of construction to the manufacturing industries, and they are the base for most productive activities.

The mining industry is an important branch of the extraction industry, based on the extraction of basic metals minerals (such as iron, copper, tin, zinc, lead, and aluminium), noble minerals (such as gold, silver, and platinum), and rare minerals (such as lithium, cobalt, and rare earth elements). The spot where their extraction is placed is denominated "*mine.*" The remaining minerals, designated as non-metallic minerals, are normally extracted from a quarry, and they can be divided into two groups: industrial and ornamental rocks.

In the first case, the rock is much less valorized than the metal minerals because it is normally abundant in nature. This type of rock constitutes the base of the industrial activities such as the production of Portland cements (clay limestone), different categories of ceramics (with diverse clays), terracotta (with the use of red clay), and glass (having sand as principal element). As it is well known, the Portland cements, on a par with the steel, are fundamental materials for the construction of buildings and heavy infrastructures. Examples are bridges, tunnels, airports, dams, and ports. The so-called construction industries are the biggest producers of waste at European level, accounting for about 34% of the total. For instance, in 2015 in Portugal, of the 145 million tons of materials extracted from the lithosphere, about 73% corresponded to non-metallic minerals, mainly extracted from quarries and not from mines. Essentially, they were materials for civil construction and public works.

The ornamental rocks, having much greater commercial value, correspond to a historically very old extraction. There are countless famous monuments of the Ancient World, some over 4000 years old, built with ornamental rocks or special limestone rocks. Ample cases are in Egypt, Greece, Minor Asia, Maghreb, Roman Empire, India, and North and Central American Indian Empires (mainly Aztecs and Mayans). Marbles, some special limestones, granites, and precious and semi-precious stones, are the main rocks used for this purpose. The possession of these minerals represents, for millennia, a strategic factor for the growth of the economy of a country and the evolution of the industrial relations (Conti, 2006).

Like mining activity, the cultivation of a quarry can be classified into three different categories:

I. **Open air**: type of quarry used to extract deposits of mineral near the surface.
II. **Underground**: it requires equipment and workers to operate under the surface of the earth.
III. **Pit**: type of open pit typical for flat areas, where mining is carried out along graded surfaces that extend downwards to below the level of the countryside (Fornaro et al., 2002).

Although the first type is the most widespread, there are examples of underground quarrying, especially for relatively high-quality materials. The Candoglia marble quarry for the Milan cathedral and numerous Carrara marble quarries are notable examples of this type of quarry (Fornaro et al., 2002).

The extractive industry is normally connected to the geological conditions of the territory where it is placed and to the location of the mineral substances. Since there is no economically viable alternative to produce these indispensable resources for our society, this situation often entails the necessity to operate the mineral in non-optimal conditions (e.g., in a difficult environmental situations or the proximity to the inhabited settlement of a city) (Brodkom, 2001). Consequently, several aspects, such as the deposit location, the mineral value, the geographic area, the transport networks, the geo-mechanical characteristics of the soil, and the social political system (Gisotti, 2008), affect the cultivation project (Fornaro et al., 2002).

According to the Italian mining law of 29 July 1927 n. 1443, the distinction between quarry and mine is very subtle. The difference essentially consists of the type of materials that are extracted and are under the management of the property. In fact, the quarries are left available to the owner of the land, while the mines are part of the state property. The areas where a quarry is implanted are directly identified, when useful economically feasible materials emerge, or they are identified through prospecting operations that interpret the nature of the subsoil. The organization of the quarry involves the construction of access roads and storm water collecting channels to avoid dangerous influxes into the quarry (Fornaro et al., 2002).

The development of the cultivation depends on the local morphology if it is in the mountains: generally, stone steps are created on the slope of the mountain and connecting roads between the mountain and the valley, consisting of inclined planes, ligature roads, and cableways. On the other hand, if the quarry is located on flat ground and the useful material does not surface, then a pit quarry is opened, proceeding first with the uncovering, or rather, with the removal of the sterile material until reaching the obtainable material. The plants that are common to all types of quarries are those related to the lines of transport of electricity, compressed air production, industrial water pipelines, explosives, fuels, lubricants, and other types of materials. As far as the machinery is concerned, in the case of a quarry on flat ground, there is a need for excavators and fixed and self-propelled conveyors; in the case of quarries in the mountains, mechanical shovels, wagons drill, drills, dumpers, conveyor belts, locomotives, and wagons are required. For the marble and ornamental stone quarries (the subject of this research), the cutting plant must be inserted with the standard helical wire (Fornaro et al., 2002).

The cultivation methods can all be traced back to the stepped type, which consists of carving, a succession of sloping platforms (steps) in the field: the variants depend not only on the type of the deposit, the mineral, the encasing rocks, and the covering but also on topographic conditions. On flat land, the one-sided method is usually used: a long step is achieved that can be parallel, fan-shaped, or block-fed. The reduction is regulated, so that the sterile can be transported on pre-established areas or unloaded opposite the front, on the area that was previously occupied by the ore that has been removed (Fornaro et al., 2002).

The access roads to the steps consist of inclined ramps with up to 20 degrees of slope, on which the machines are moved to reach the workstations, while the excavators on belts, dumper trucks, or railway carriages discharge the sterile. In the mountains, the methods with open steps are used, which follow the slope according to the level of the curves, and a funnel, in which the steps follow closed bends that are ever narrower towards the bottom. In the second case, a jet cooker opens at the bottom, where the mined mineral is conveyed and arrives either on tapes or other means of transport to go to an extraction well. For piercing siliceous rocks, jet piercing is often used; the flame dart of an oil-oxygen mixture burner replaces the mines. In the quarries of marble and other cutting stones, as already mentioned above, we continue with the use of the helical wire with which large blocks from the living rock are cut (Fornaro et al., 2002). Currently, on the European and global stage, the mining industry related to marble is faltering.

1.1.1.2 Undesirable Effects of the Extraction Industry

Most often, due to the crisis of the building sector and other economic and geological aspects, a considerable number of quarries is subjected to abandonment or closure. Vulnerable to a brutal materials removal, the quarries are forced to see their state completely changed, and there is no hope for them to return to their original condition. For these reasons, they are definitely considered the symbol of the gap, turning them into brownfields (Talento et al., 2019). The brownfields are places of degradation, previously exploited for industrial scopes, that they can be considered as a depreciation of the landscape and an ecological, aesthetic, and social loss (Dumesnil & Ouellet, 2002). Many other terms can be used to define the residual and abandonment spots, but the most important are “Friche” (Chaline, 1999), “Terrain Vague” (Barron, 2013), “Wasteland” (Nabarro & Richards, 1980), “Drosscape” (Berger, 2007), “Derelict Land” (Kivell & Hatfield, 1998), “Terra incognita” (Bowman & Pagano, 2004), “The third landscape” (Clement, 2003), “Zero panorama” (Smithson, 1996), “Urban sinks” (Lynch, 1990), “Dead zones” (Doron, 2000), “Il vuoto” (Secchi, 1989), and “In between spaces,” “White areas,” “Blank areas,” and “SLOAPs” (Spaces Left Over After Planning) (Jonas & Rahamann, 2015).

Despite the abundance of definitions, these discarded landscapes are gathered to a unique aspect, namely, that of having been expelled from the cities for their incompatibility with the urban phenomena. This incompatibility is represented by the wastes from which the abandoned landscapes are characterized. For this reason, one of the most brutal consequences produced by the extraction activity is, undoubtedly, the destruction of the landscape in its geographical and identity sense. Indeed, this type of industry is an important form of degradation of the territory made by man, so far as to alter or even to break the fragile natural equilibrium achieved over thousands of years (Muzzi & Rossi, 2003). The presence of a quarry or a system of quarries on the territory considerably affects the ecological, hydrological, and pedological balance, provoking an irreversible mutation of the territorial context, such as the loss of the high-quality components of the landscape (Labianca, 2013). From these premises derives the urgency of evaluating the alterations produced by the quarries, not only in the aesthetic field but also considering a large range of effects provoked in the entire pre-existing ecosystem.

1.1.2 Research Proposal

Research intention and application.

The proposed study aims to further investigate, from the architectural and landscape point of view, a different methodology of requalification for the industrial abandoned sites, such as the quarries. This systematic methodology, described in Chap. 5, includes the definition and the elaboration of scientific criteria, a formula of practical use, and the derivation of main guidelines for the rehabilitation of an extractive site. Based on this systematic methodology, the rehabilitation of an extractive site implies not only the reconversion of the pit but also the reuse of the wastes resulting from the extraction activity. Involving the wastes in the process of rehabilitation is extremely important to preserve the identity of the place and its historical and industrial memory and to improve sustainability, applying the concept of circular economy.

1.1.3 Study Sphere

Research dimensions, from broad to specific field of the study.

The research, jointly with its methodological process, has an interdisciplinary approach. It covers the following disciplines: **Planning**, **Landscape,** and **Architecture**, within a geological context.

Study Area Estremoz Anticline, Portugal.

Object of the Study System composed by marble quarries and waste heaps, namely, "*escombreiras*".

Research Line Rehabilitation of the industrial landscape of the Estremoz Anticline through an operation of replanning of the abandoned extractive site (based on the elaboration of a new systematic methodology), the rehabilitation of quarries, and the reuse of waste heaps in landscape and architecture.

1.1.4 Research Field

Research delimiting boundaries.

The object of the study, explored in this book, is represented by the system of quarries (with the corresponding waste heaps) located in the Estremoz Anticline, in Portugal. For this purpose, several quarries have been analysed to define new rehabilitation criteria for the valorization of the abandoned industrial area. Due to reasons of physical and regulatory incompatibility, the current active quarries will be shown in this selection, but they will not be considered for design and research scopes.

1.1.5 Problematics

Statement of the problems that prompted the decision to further the initial investigation.

The beginning of any research carries with it some problems of approaching the topic. Most often these problems represent the instrument that motivates the research and its development. Related to the current work and having it as target a massive area of about 280 km^2, one of the main problematics is to deal with the large scale. The following concern:

I. The difficulty to have an interdisciplinary view of the research confronting, at the same time, different disciplines.
II. The effort to deal with a huge amount of waste material and so many quarries, located in the study-area.
III. The trouble to glimpse beauty in a degraded industrial landscape.
IV. The problem to think an integrated architectural and landscape solution for the rehabilitation of the abandoned area that does not interfere with the industrial activity, still in progress.

1.1.6 Research Questions and Hypothesis

Proposed theory and queries that will assist in its confirmation.

To regenerate an abandoned industrial area with an operation of replanning of the territory, the rehabilitation of quarries and the reuse of waste materials means that, when one analyses these unusual elements, many questions remain unanswered. Questions can be classified according to planning, architectural, and landscape aspects. Those aspects have great influence on the concept of the rehabilitation of a quarry and the re-valorization of the waste landscape, which is the main interest of the research.

The general **planning** question, posed by the research, can be resumed as follows:

I. *Will it be achievable to demonstrate the effectiveness of transforming a disused industrial landscape into an attractive, sustainable, and visited place, through an operation of replanning?*

Furthermore, the main **landscape** question posed by the research is the following:

I. *Will it be achievable, through the elaboration of innovative sustainable strategies, to define the best form of rehabilitation of each landscape unity of the Estremoz Anticline?*

The landscape field also contains two additional secondary questions:

2. Will it be conceivable to confer the traditional features of a landscape, such as beauty, equilibrium, harmony, and functionality, also to the degraded, abandoned, and wasted landscape?
3. Will it be feasible to prove that the quarries and the huge accumulations of wastes will no longer be a burden, with a negative impact, on the landscape?

The main **architectural** question formulated by this research is the following:

I. *Will it be feasible to involve the waste heaps in the process of rehabilitation of the area, creating original buildings and public spaces from them?*

The architectural domain also includes two other secondary questions:

1. Will it be possible to redevelop the "stone landscape" using existing forms and materials (quarries and mounds of marble wastes) to preserve the historical and the industrial memory of the place by means of a policy of the waste reuse?
2. Will it be possible to make compatible the artificial elements with the natural ones by studying and creating relations between waste heaps, quarries, and the surrounding landscape?

Hypothesis

The research hypothesis is based on two important and central arguments to be confirmed by the study. The first one is represented by **the concept that suggests acting exclusively in situ**, therefore, not exporting any kind of material. It aims to leave everything as it is, but looking at the discarded material with different eyes, seeing the *escombreiras* not as imperfections to be either suppressed or hidden but as elements having good potential for the redevelopment of the industrial area, having also its own entity and utility.

They are "*traces deriving from the action of man that are transformed into signals and overlap with the genesis of the territory. To re-evaluate a site it is necessary to consider these signals, understanding the characteristics of the site and making the new system compatible with the existing one*" (PROAP, 2010).

The developed research is based on the concept of giving a new meaning to what is ordinary, working with the wastes rather than excluding them, a priori, to preserve the historical and industrial memory of the place, not removing marble wastes from the context they belong (Cillo, 2008; Clement, 2005, 2011; Clement & Lucchesini, 2013). In this way, the *escombreiras* do not represent disturbing factors of the landscape but iconic elements of the stone territory because they provide a dynamic image, coherent and aligned with the context and the concept of the landscape itself (Hichours et al., 2007; Kastner, 2010).

The second main argument is characterized by the interesting **binomial "architecture-landscape"** which represents the essence of the current research. Due to the complexity and interdisciplinarity of the topic, it is impossible to think of the two elements of the binomial individually. On the landscape side, the topic will be verified through the study of the evolution of the landscape, the analysis and description of the abandoned one, and the replanning of the territory under investigation. On the architectural side, the topic will be confirmed by analysing and categorising existing case studies of quarry regeneration and elaborating original sustainable strategies for the rehabilitation of quarries and the reuse of waste heaps, considering them as containers of spaces with existing low-cost technologies and materials.

1.1.7 Research Objectives

Goals to be achieved with the proposed investigation.

The proposed research has general and specific objectives, as described below.

General Objective

I. The main objective of the research is to re-qualify the extractive landscape of the Estremoz Anticline with elements and materials available in situ, to preserve the historical and industrial memory of the place. As will be shown in Chap. 5, a new systematic architectural methodology will be proposed to demonstrate the possibility of transforming an abandoned industrial landscape into an attractive and visited one, having an associated aesthetic value.

The specific objectives can be classified according to the planning, architectural, and landscape aspects, and they are conforming to the research questions previously described.

Planning Objective

I. The research, before presenting an architectural and landscape proposal for the practical rehabilitation of a specific subarea of the Estremoz Anticline, aims to think on the macro scale through an urban replanning of the complete Anticline, imagining new macro-uses for the abandoned areas of the extractive industry that they do not conflict with all those where there are still active quarries.

Landscape Objective

I. From the general planning of the Estremoz Anticline, the research, with the elaboration of innovative sustainable strategies, will define the perfect form of regeneration of each landscape unity, conferring beauty, harmony, and functionality to the abandoned industrial landscape and proving that the huge accumulations of waste will no longer be a burden, with negative impact, on the landscape.

Architectual Objective

I. The study, after choosing a small specific subarea of the Estremoz Anticline and making a drone survey, intends to put into practice the concept of waste reuse, through the proposal of architectures and public spaces built inside the mounds of wastes, linked to the corresponding quarries and the surrounding landscape.

1.1.8 Research Methodology

Method of investigation applied to the object of study.

The first step of the investigation consists of searching and selecting various types of documents and books and organizing an extensive bibliography research for the drafting of the state of the art. Its scope is to provide, in a transparent and methodized form, an exhaustive report regarding the landscape concept, the rehabilitation of quarries, and the notion of waste reuse in landscape and architecture. This step is extremely important to setting up the bases for the rest of the research.

Several references have been selected for the organization of the literature, and they have been searched for keywords, mostly related to the concepts of landscape, industrial landscape, drosscape, quarry rehabilitation, waste, waste heaps, waste reuse, recycle, resource recovery, circular economy, sustainability, and architecture. The bibliography has been then organized following the intention of the research, namely, that of creating a fluent reading about the waste notion, from its origins (also evaluating the change of the landscape and example of quarry rehabilitation) to its contemporary mutations.

From the discussion of the theory of the landscape, quarry regeneration, waste reuse, and circular economy, the research provides a critical survey about some practical examples of reuse, applied to quarries and the dumped material, both in landscape and architectural fields. This combination of theory and practice contributes to the elaboration of new typologies of reuse for the rehabilitation of quarries and waste heaps. As will be described in Chap. 5, these typologies of reuse, associated to the features of each quarry and waste heap, allow to obtain new criteria for the reconversion of an extractive site (primary guidelines for planners for the rehabilitation of an extractive site located in any place with whatever features) and a scientific formula of practical use, filling an existing gap in both architectural and theoretical domains. In this way, a new systematic methodology has been created for the rehabilitation of an abandoned extractive site, which involves the reuse of the discarded material.

To the best of the author's knowledge, there is not any similar schematization in the literature, and the idea of dealing with the mounds of wastes, in the architectural and landscape sectors, has been slightly approached until now. On the contrary, the work presented in this research forms the basis to comprehend how the reuse idea can advance in the architectural discipline, following a new systematic methodology, which is extremely crucial for the replanning of the industrial area of the Estremoz Anticline.

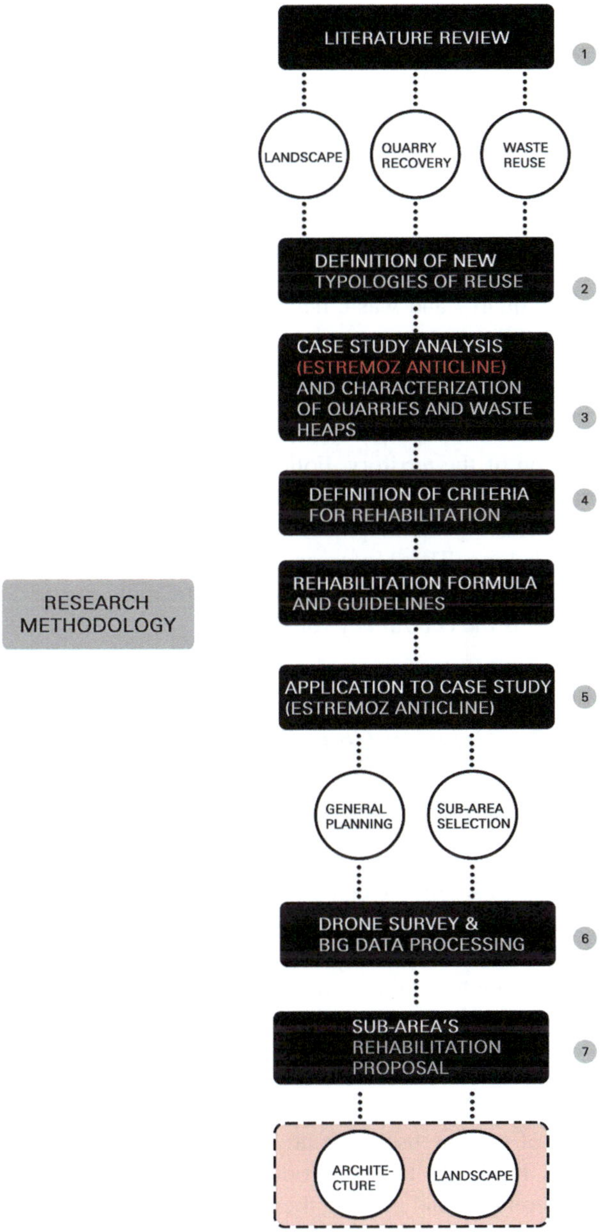

Fig. 1.1 Block diagram resuming the proposed research methodology. (Katia Talento, 2021)

The following developments of the methodology (resumed in Fig. 1.1) are related to the practical part of the research, and they start from the material identification of the area on the territorial maps. Through this identification, one realizes the large size of the area (about of 280 km^2) and the significant number of quarries and waste heaps located on it (nearly 476). As it will be shown in Chap. 5, after having characterized every single quarry and waste heap belong to this massive area, conferring them specific parameters with the use of ArcGIS™ Software, an operation of replanning of the Estremoz Anticline has been developed, according to the criteria previously elaborated. In this way, new macro uses have been allocated to the general area for the rehabilitation of the industrial abandoned territory.

As described in Chap. 5, the following steps of the research methodology pay attention to the architectural field and the proposal of an architectural and landscape solution for the application of the waste reuse notion in the practical sector. This solution has been designed in a specific subarea of the Estremoz Anticline for a given quarry and waste heap. It represents the culmination of the research, and it physically demonstrates how an abandoned quarry can be regenerated, including the associated waste heap in its process of regeneration.

To make this feasible, it is necessary to select a small intervention subarea and to carry out a topographic deepening through the realization of a drone survey of this portion of the territory. Following this research methodology, the work intends to confer a new meaning to what is ordinary, working with waste to preserve the industrial memory of the place, not exporting marble residues from the context where they belong. These interventions are blended in the surrounding territory, and they connect the architecture discipline with landscape and planning, using interdisciplinarity as the central key for change.

1.1.9 Original Contributions

Major original contribution.

From the author's point of view, the major original contribution, described in this book, deals with the elaboration of criteria and typologies of reuse for quarries and waste heaps, to formulate a systematic architectural methodology for their rehabilitation. The work, for defining the referred criteria, is fully supported by a unique and complete database created by the author.

Articles and papers published in the context of the present research.

The main contributions of the work, described in this research, extend from the analysis of the landscape changes to the rehabilitation of quarries and from the concept of waste reuse to an architectural proposal about this notion. These contributions have led to the production of several papers in scientific journals and conferences, and each of them represents a specific phase of the research process. The contributions can be listed as follows:

Journal Articles

- Landscape—A review with a European Perspective. *Land* 2019, 8(6), 85; https://doi.org/10.3390/land8060085
- Eco-friendly reuse of marble wastes in landscape and architecture. *WIT Transactions on Ecology and the Environment* 2019, 238, 665–677. Paper DOI: 10.2495/SC190571

- Quarries: From Abandoned to Renewed Places. *Land* 2020, 9, 136. https://doi.org/10.3390/land9050136
- The Reuse of Waste Heaps from Extraction Sites: An Architectural Methodology. *Sustainability* 2020, 12, 6548. https://doi.org/10.3390/su12166548
- The reuse of marble waste heaps in the Estremoz Anticlinal. *Identidades: territorio, cultura, patrimonio*, 2020, 9. Paper DOI: 10.5821/identidades.10266.

Conference Papers

- The Metamorphosis of the Landscape: The Dynamic Reuse of Marble Waste. *The fifth International Conference on Architecture and Built Environment with AWARDs S. ARCH*, 22–24 May 2018, Venice. Book of Abstracts. Get It Published. ISBN 978-3-9818275-7-6.
- The Metamorphosis of the Landscape: The Adaptive Reuse of Marble Waste. *The fifth International Conference on Architecture and Built Environment with AWARDs S. ARCH*, 22–24 May 2018, Venice. Conference Proceeding. Get It Published. ISBN 978-3-9818275-9-0.
- Eco-friendly reuse of marble wastes in landscape and architecture. *The Sustainable City XIII*, 1–3 October 2019, Valencia.
- The reuse of marble waste heaps in the Estremoz Anticlinal. *XII SIIU, Seminário Internacional de Investigação em Urbanismo*. 25–26 June 2020 Lisbon.

Bibliography

Barron, P. (2013). *Terrain vague: Interstices at the edge of the pale*. Routledge.

Berger, A. (2007). *Drosscape. Wasting land in urban America*. Princeton Architectural Press.

Barker associates. (n.d.). *Sustainable architecture: What is it and How do we achieve It?* Barker Associates.

Boca, D., & Oneto, G. (1989). *Discariche, cave, miniere ed aree difficili o inquinate*. Il Sole 24 Ore Pirola.

Bowman, A., & Pagano, M. (2004). *Terra incognita: Vacant land and urban strategies*. Georgetown University Press.

Brodkom, F. (2001). *Codice di buona pratica ambientale nell'Industria Estrattiva Europea*. PEI srl.

Budd, M. (2002). *The aesthetic appreciation of nature: Essays on the aesthetics of nature*. Clarendon Press.

Chaline, C. (1999). *La régénération urbaine*. Presses Universitaires de France.

Cillo, B. (2008). *Nuovi orizzonti del Paesaggio*. Alinea.

Cipriani, L. (2012). *Vita e morte del paesaggio industriale*. Università degli Studi di Trento.

Clement, G. (2003). Le manifeste du Tiers-Paysage. Editions Sujet/Objet.

Clement, G. (2005). *Manifesto del terzo paesaggio*. Quodlibet.

Clement, G. (2011). *Il giardino in movimento*. Quodlibet.

Clement, G., & Lucchesini, G. (2013). *Giardini, paesaggio e genio naturale*. Quodlibet.

Conti, S. (2006). La produzione mineraria ed energetica. In *Geografia dell'economia mondiale* (pp. 123–156). Utet.

Doron, G. (2000). The dead zone and the architecture of transgression. City: Analysis of urban trends. *Culture, Theory, Policy, Action, 4*(2), 247–263.
Dumesnil, F., & Ouellet, C. (2002). La réhabilitation des friches industrielles: Un pas vers la ville viable? *Vertigo La Revue Électronique En Sciences de l'environnement, 3*(2) http://vertigo.revues.org/3813
Fornaro, M., Lovera, E., & Sacerdote, I. (2002). La coltivazione delle cave e il recupero ambientale. In *Estrazione di materiali per usi industriali e pietre ornamentali* (Vol. 1). Politeko.
Frare, G. P. (1996). Cave. Tecniche di coltivazione e recupero ambientale. Zoppelli.
Gerbella, L. (1960). *Arte Mineraria*. Hoepli.
Gisotti, G. (2008). Le cave. Recupero e pianificazione ambientale. Flaccovio Dario.
Hichours, Y., Hourimche, A., Benbouziane, M., Tahiri, O., & Ejjaouani, H. (2007). Interests and environmental impacts of the quarries of exploitation of the region of Casablanca. *Moroccan Magazine of the Civil Engineering, 120*, 8.
Jonas, M., & Rahamann, H. (2015). Tokyo void: Possibilities in absence. Jovis Edition.
Kastner, J. (2010). *Land and environmental art*. Phaidon.
Kivell, P., & Hatfield, S. (1998). Derelict land—Some positive perspectives. In *Environment, planning and land use* (pp. 118–129). Ashgate.
Labianca, M. (2013). La riqualificazione dei siti estrattivi: Un'opportunità per il territorio. La situazione della Puglia. *Geography of Italy, 120*, 55–70.
Lynch, K. (1990). *Wasting Away*. Sierra Club Books.
Muzzi, E., & Rossi, G. (2003). *Il recupero e la riqualificazione ambientale delle cave in Emilia-Romagna*. Arketipa.
Nabarro, R., & Richards, D. (1980). Wasteland: A Thames television report. Thames.
ONU. (2016). *Guia sobre Desenvolvimento Sustentável*.
PROAP. (2010). *Landscape architecture*. Note Editions.
Secchi, B. (1989). II vuoto. Un progetto per L'urbanistica. Einaudi.
Smithson, R. (1996). Sites and settings. In *Robert Smithson: The collected writings*. University of California Press.
Talento, K., Miguel, A., & Kullberg, J. C. (2018). The metamorphosis of the landscape: The adaptive reuse of marble waste. In The 5th international Conference on architecture and built environment with AWARDs S. ARCH 2018 CONFERENCE PROCEEDING (Get It Published).
Talento, K., Amado, M., & Kullberg, J. C. (2019). Landscape—A review with a European perspective. *Land, 8*(85). https://doi.org/10.3390/land8060085
Vallario, A. (1998). *Attività estrattive, cave e recupero ambientale*. Liguori.
Veríssimo Serrão, A. (2011). *Filosofia da Paisagem, Uma Antologia* (2nd ed.). Centro de Filosofia Universidade de Lisboa.
WMO. (1985). *A report of the international conference on the assessment of carbon dioxide and other greenhouse gases in climate variations and associated impacts,* (No. 661).
World Commission on Environment and & Development. (1987). *Our common future*.

Part II

Chapter 2
Landscape Mutation

There are places. Abandoned places. Forgotten places. These are the places I like to be.

Dawn Kurtagich

2.1 Introduction

Analysis of the landscape from its origins to its contemporary mutation, caused by the strong industrialization and the strong brand of man on the natural territory.

Our era is decisive of a popularized and democratized landscape that belongs to everyone, while, in the past, it had the role of a social code and a distinctive sign of an elite that recognized itself in the common sharing of representative places or topical representations.

The landscape has passed from a peripheral vision to a central one, in which it has become important in several disciplines, especially in philosophy and geography (Jakob, 2009; Raffestin, 1978). The reasons for the success of landscape, at first sight, could refer to the ecological movement. The ecological and landscape consciousness (or even: the desire of nature and the desire of landscape) are both part of a wider current, which refers to the relationship between post-modern man and nature. The ecological movement of the early 1970s represents, from this point of view, a remake (Fedeli, 1990; McHarg, 1969), due to the fact that the malaise in the city and urban crisis generated the demand for more nature. However, there are also two other significant aspects that have led to this change.

The first one concerns the crisis of post-war planning and the growing dissolution of habitual divisions between the identifiable poles of the territorial system: city-countryside-nature and city-industry-countryside-nature. The known territories were transformed into an increasingly unreadable reality due to the "sprawl" phenomenon (Ingersoll, 2006; Berger, 2007), the urbanization of small towns, the crisis of agriculture (45), and traditional industrial activities followed by an exponential birth of urban fringes. The lack of references and daily comparison of non-places

K. Talento, *Sustainable Buildings from Marble Waste Heaps*,
https://doi.org/10.1007/978-3-031-72146-5_2

and other interstitial spaces (46) has motivated the desire to identify and protect the landscape. This necessity gave the beginning to firmly speaking about this theme, trying to expose it, and give it an immobile form in the texts, inscribing in the "Great Book" of identified sites as the universal heritage of humanity; also, the UNESCO activity has undoubtedly contributed to making the landscape popular.

The second aspect relates to the impacts of our images of civilization on the landscape. On one hand, we are in fact surrounded by countless landscape images that reach us from screens, advertising panels, or newspapers. On the other hand, there are many landscape images that are created by us while traveling. Nowadays, the world of touristic industry and the digital image industry have a huge impact on our way to discover and memorize the landscapes. Feeling completely free in the enjoyment of very different landscapes, we are, in reality, conditioned by a cultural and economic device; the landscape seems to correspond to the idea of an aesthetic value that is imposed by others without us being aware of it (Jakob, 2009).

This chapter provides an exhaustive report regarding the landscape concept, organizing an extensive bibliography research to understand, for the first time and in a single text, the flow of the perpetual evolution of the landscape. In fact, in the present chapter, it is possible to read the entire way of this mutation, since the origins of the landscape to its current change, which is essentially caused by a strong human influence due to economic and industrial reasons. The work describes distinct views of the natural territory, showing its strongly humanized and altered side due to the presence of many industries that are located on the land; especially, the scenario of the abandonment is treated, which is formed by several industrial wastes. The concept of recycling, through the promotion of the landscape architecture methodologies for the rehabilitation of the devastated territories, is also addressed, and it tries to highlight their potential of re-use for a different requalification of the lands.

When considering that this natural scenario has always been distinguished by beauty, equilibrium, harmony, and functionality, will it be possible to also confer these characteristics to the degraded, abandoned, and wasted landscape? This is the key question that motivates the investigation reported by the study, to make understand that human action is not always a negative factor. At the same time, new ways of requalification and research exist, to define the recovery and rehabilitation approaches of the landscape.

Section 2.2 discusses the landscape concept from its birth, its origin, the etymology of the word "landscape," and some theories from different disciplines about the concept, and it discusses the legislation and the concept of identity. Section 2.3 describes the landscape mutation, analyzing abandonment and residual spaces, such as *friches*, *terrain vagues*, and drosscapes while paying attention to the case of the quarries. Section 2.4 shows the redesign of the degraded landscape that deals with the concept of waste reuse, the rehabilitation concept as a bridge between the past and present, and the promotion of art and landscape architecture in abandoned territory. Finally, Sect. 2.5 draws the main conclusions.

2.2 The Concept of Landscape—Birth and Definition

2.2.1 Origin and Etymology of the Word "Landscape"

The origin of the term landscape is relatively recent, and it separates the European attitude in two periods of time: pre- and post-landscape. The concept of landscape, as it is understood nowadays, did not exist; society was more oriented toward the interior, almost exclusively extending the observation of nature to the cloister or vegetable garden, since the external landscape was only a background for other scenes (O. Ribeiro & Lautensach, 2009). The gaze on the world has been externalized over the centuries, as long as man exercised a growing dominion over the environment that surrounded him. The first occurrences of this term are found in the second half of the fifteenth century with the introduction of the landscape element in painting, one of the innovative topics of the Renaissance (Veríssimo Serrão, 2011).

At that time, the landscape was always seen as a marginal element: the nature was represented in a schematic and essential way, only to propose the settings with sacred stories. The naturalistic element was progressively reborn during the Middle Ages; the main founder of this revolution was Giotto, who, in the impressive series of frescos made in the Basilica of Assisi at the end of 1200, for the first time reproduced a real landscape within painted stories. Giotto's characters moved in one realistic space, against the background of cities, medieval villages, and rocky or wooded landscapes that the observers could recognize (Fig. 2.1) (Salvini, 1964).

Between the end of the fourteenth and the beginning of the fifteenth century, during the spread of the "late Gothic" style, the landscape representation began to change with the introduction of the detailed and meticulous illustration of every single natural element. The circulation of manuals, which is called "*Tacuina Sanitatis*," containing studies and drawings of plants and animals, reinforced the interest in the landscape. These texts were used for medical and culinary purposes, which contributed to the spread of reproductions of herbs and plants illustrated with naturalist precision (Fig. 2.2) (Hanley, 2013). It was only during the Renaissance age that painters began to study the landscape and the nature with scientific precision. Particular attention should be dedicated to the pieces of Leonardo da Vinci, who marked a fundamental transition in the evolution of the landscape painting. Leonardo performed a lot of studies "*dal vero*" (Fig. 2.3) representing natural phenomenon that helped him to recreate landscapes with incredible realism (Fig. 2.4). After this period, little by little, the landscape painting became entirely protagonist, and it obtained the recognition of an autonomous genre unbound by the presence of human (Cricco & Di Teodoro, 2017).

This brief summary of the painted landscape history illustrates that it took several centuries in order for the nature to be represented as an object of pleasure worthy of attention (Jakob, 2009). The landscape has passed from a passive perspective to a more active one developing itself, even in the architecture field. In Great Britain, it has been the first example of a new landscape architecture school, which perceived and realized proposals to intervene in the landscape, creating parks and

Fig. 2.1 Giotto—Legend of St Francis—The gift of the mantle (1294) (Wikipedia, 2019)

gardens that are inspired by idealized rural landscapes (Jellicoe & Jellicoe, 1975). At the end of the twentieth century, the landscape became fashionable, and there was a substantial curiosity regarding this concept. The landscape appeared to be the key to responding to many questions regarding the management of space and the activities that were developed within it, due to its potential for integration (Pinto-Correia, 2000).

In this same period, even arts began to take landscape into consideration, as we will further observe in the development of the contemporary artistic movement that dealt with the landscape in a new and unconventional way. Following the ideas deriving from the Dadaist vanguard, this new movement directly acts on the landscape, modifying it, permanently or temporarily, with the introduction of enormous installations. These pieces are the result of creative and artistic choices that place the real landscape as the main subject (Tufnell, 2006; Weilacher, 1999). If one wants to rest on the meaning of the word landscape, one can state that there are several

Fig. 2.2 Tacuina Sanitatis—Mandrake Dog (1389) (Wikipedia, 2019)

research studies regarding the origin of it. However, among the various interpretative strands, what seems to offer greater levels of depth and further developments is the one that was presented by the French semiologist Jeanne Martinet, who dealt with the theme at a conference held at the University of Saint-Ètienne in 1983. According to Martinet, the concept of landscape derives from the Dutch word "*landschap*" which was used by Dutch painters during the second half of the fifteenth century to shift the interest from the representation to the represented model. In their pieces, the nature becomes the subject of the painting with a new approach, which, based on the words of the painter Joachim Paternier, "*it reverses the scale of*

Fig. 2.3 Leonardo da Vinci—Study of a Tuscan Landscape (1473) (Wikipedia, 2019)

values recognized until that moment and it extends the landscape to the detriment of the figures hold as subordinates" (Martinet, 1984).

Yves Luginbühl, a French agronomist and geographer, as well as one of the main founders of the European Landscape Convention, has a different vision regarding it, although he confirms the Dutch origin of the word. He believes that the matrix is composed of "*lant*" (territory) and "*scap*," equivalent of the German term "*schaft*" (community); it is, therefore, a word that links the territory to the community, born in the Dutch, Danish, and German coastal territories in which the respective local people were settled. They were marshy lands at the edge of the North Sea, which were inhabited due to terpen and embankments obtained by accumulating the surrounding lands to provide shelter from the tides (Luginbühl, 2009, 2017).

The connection between territory and community, which constitutes the Dutch matrix of landscape concept, is confirmed by the French word "*Paysage*," which was used in the middle of the sixteenth century, destined to act as the main vehicle for the dissemination of neo-Latin side of the European languages ("*Paesaggio*" in Italian). Jeanne Martinet also notices the work of etymologists, whose outcome consisted of determining that "*pays*" derives from Latin "*pagensis*," which means both an inhabitant of a "*pagus*" (village) and the territory of a village. As for the suffix-age, the same author does not exclude, despite general rules, the sense that indicates the action or the result of an action, even if it is applied to a noun and not a verb. Therefore, the term landscape would mean "making a country," the expression of a place, a specific representation of the relation between the community and its habitat, at the same time allowing for the global apprehension of a place (Cagnato, 2016).

Fig. 2.4 Leonardo da Vinci—Madonna of the Yarnwinder (1501) (Wikipedia, 2019)

2.2.2 Selected Landscape Theories

Based on the observations above, it is possible to affirm that the word "landscape" is a rich and complex word that often correctly represents different and contradictory concepts. There have been many attempts at its interpretation, often pertinent, but it has never been possible to reach a shared formula that contains all of the facets composing it (Cagnato, 2016). In this way, the scientific debate regarding the concept of landscape is always open to many inherent disciplines (geography, geology, engineering, architecture, urban planning, agronomy, land art, etc.), when considering the high fluidity and richness of facets of the topic that has never had a precise disciplinary collocation in an academic sense. The consequence is the presence of various theories about this broad and interdisciplinary theme. James Corner, landscape architect and theorist, sustains that there is one apparent topic that unifies all the fields of landscape, elaborating a single theory: the landscape imagination. According to his thesis, there is nothing natural about landscape: even though landscape invokes nature and engages natural processes over time, it is first a cultural construct, existing as a product of imagination. Landscapes are initially imagined and represented in images, words, and projects afterwards. Like music, landscapes are forged by the imagination, while also themselves provoking and stimulating the imagination to see and figure out new things (Corner, 2014).

Pursuant to this theory, landscape is always seen as the result of an architectural project that stimulates the mind. It is wrong to make the landscape coincide with nature, because, like a garden or a beach, it is not a landscape but, simply, an already existing scenario that is not linked to any architectural project. On the contrary, the philosopher George Simmel states that the landscape exactly derives from nature, being a part of nature itself [55], an attractive nature that sensibly shows itself in the aesthetic experience. In this way, the thought of the French philosopher Paulhan is also aligned. He states:

> We like the landscape because we like nature, and the landscape is the portrait of it. (F. Paulhan, 2016, p.5).

Differently, the verses of François Amiel (Amiel, 1931), "*Le paysage est un état de l' âme*" (landscape is a state of the soul), suggest a romantic vision of the landscape and the attention toward the interior landscape. It reflects the gaze on the world of everyone, a purely subjective vision that is linked to existence, to memories and emotions that are connected to the landscape. Our past is shaped by the constant presence of that landscape, being formed by people, things, and imagination, always alive in the unfolding of the experience. The emotional bond between people is certainly crucial and indispensable, but the relationship with the significant entities of the landscape is also essential: the horizon of the sea, the smell of a neighborhood, or a particularly significant road. The investigation of the inner landscape aims to analyze those deep bonds that intimately unite places with personality and experience. This concept is translatable with the Anglo-Saxon term "Inscape" (intern point of view), which was used for the first time by the Irish poet Gerard

Manley Hopkins. Using this term, he defines that complex of characteristics that gives uniqueness and exclusivity to an individual experience, resulting in being different from any other experience (Andreotti, 1996, 1998; Lettini & Maffei, 1999). This psychological attitude finds its extreme level in Fernando Pessoa's (Soares, 1982) "*Livro do Desassossego*" (The Book of Disquietude)—"The landscape is a state of soul," with the difference that the landscape is not an affective projection on the outside but, conversely, a personal picture in which an individual subject receives, as on a screen, the world projected on him.

Talking about the landscape, one finds another important theory that is related to the description of the cultural landscape and created by the German scientist Alexander von Humboldt. Based on his doctrine, the landscape that is observed by man is the result of a complex and dynamic system, where varied natural and cultural factors are integrated influencing each other and changing over time. The cultural landscape is generally understood as one that man has transformed, directly or indirectly. Humboldt sustains that, in addition to the reciprocal relationship that is established between a territory and its inhabitants, it is the fact that a human being observes what surrounds him and based on that builds an image that transforms the environment or territory into a landscape (Jones, 1991, 2003; Twigger-Ross & Uzzell, 1996). As the Belgian geographer Marc Antrop explains, the importance of perception is intimately related to a holistic approach that considers the landscape as a whole is more than the sum of the parts as being the necessary synthesis for the true understanding of this whole (Antrop, 2000).

Additionally, the landscape represents a central conception and a summary resulting from a long relation between natural conditions and human action in classic geography. This relation is organized into societies, bearers of history, culture, and technological evolution. Rather than morphology, geographical landscapes contain an anthropological depth, a revealing remembrance of numerous sediments or signs that are left by subsequent transformations. Landscapes are cultural heritage, comprising an essential element of the identity of people or even a model of nation–state cohesion. Moreover, ideologically, the regions, areas of extension of unity, and models of the landscape have legitimized the organic theses of the equilibrium of the nation in its diversity and relationship between environments and ways of life, and the stable picture of the mosaic of the typical landscapes of each region. The study of the landscape is almost a seduction exercise: dismounting to perceive, finding the marks of time, the vicissitudes of history, seasons, fields, forms of settlement, buildings, and materials. More than just a concept, the study of landscapes identifies itself with the object of study of geography, the distinctive element of a clear disciplinary positioning between the natural and the human sciences (Domingues, 2001).

The theory of an Italian philosopher, Rosario Assunto, is also included in this area, and it describes the landscape, starting from some geographical elements. In his opinion, the definition of landscape can be enunciated, starting from the clarification of the meanings of three essential words: landscape, territory, and environment, which today are used almost interchangeably. However, the terms are quite distinct: "Territory" has an exclusively spatial meaning; in fact, it is roughly understood as an extension of the earth's surface that can be delimited according to geophysical divisions or political-administrative delimitations (substantially territory is matter). On the other hand, the concept of "environment" has two meanings: a biological one, which refers to the conditions of physical life, and another historically cultural one that refers to customs and traditions. Substantially, it is the content that includes the "territory" in itself because there cannot be an environment without territory. The definition of "landscape" spontaneously emerges as a form that the environment confers on the territory or, rather, the form in which the synthetic unit of matter (territory) and the function of content (environment) are expressed (Assunto, 1976). Substantially, the landscape for the author is "*the country considered from the artistic point of view*" (Assunto, 2006).

However, the landscape is not just a physical element; in fact, it exists because there are those who look at it, giving it meaning. It is an encyclopedia whose reading capacity depends on the cultural level of the observer, as an anthropological interpretation of socio-ecological facts and processes: environment, history, and geography (Turri, 2003). Consequently, the landscape is the artificial result of a culture that perpetually redefines its relationship with nature, in which the subject is entirely part of it. In this way, we witness the birth of the social factor of the landscape, which begins with humans whose activity leaves traces on the territory (Tilley, 1994). In this regard, in the famous "Social Formation and Symbolic Landscape" (Cosgrove, 1998), the author Denis Cosgrove, an American geography professor, describes the two possible attitudes that man implements toward the landscape, and he leads them to two ideal typical figures, the "Insider" and the "Outsider." The insider is the rooted inhabitant who has knowledge of places based on the memory that is transmitted by the group and on his personal memory; he produces the landscape by living it and working in it. On the other hand, the outsider, such as a tourist, is an external and disinterested spectator who has knowledge of places based on instant fruition, predominantly aesthetic; it does not produce landscape; if anything, it consumes it. For the two groups, the landscape has two different functions: for the first group, it is the transmission of identity, and for the second group, it is the recreation. Nevertheless, it is crucial to answer who, in our society, is a rooted inhabitant and who is only a disinterested tourist. We are all insiders and outsiders for a network of places. Perhaps, according to the words of the European Convention, we could talk about multiple "life frameworks." We are never completely unrelated to a place.

Other authors, such as Daniela Poli, have reported the attention to different ways of relating to places, when compared to the insider and the outsider. For example, a "returning resident," a person who, although living elsewhere, returns for a period to his place of origin having a different gaze from that of the inhabitants, consequently perceiving resources and issues with more clarity and maintaining an emotional relationship. This is the figure that Daniela Poli (Poli, 2002) has described as a "caretaker," a person who takes care of the places by choice, an attitude that may concern, for example, the chosen place for leisure or rest. Also pushed by ecologism, the question is directed toward a global sense of places, with the invitation to consider Earth as a single house that must be taken care of or even as a single garden, of which the man is the gardener (Clement, 2008), which highlights the global sense of the landscape:

> Only when I saw the Earth from space, in all its ineffable beauty and fragility, I realized that the most urgent task for humanity is to take care of it and preserve it for future generations. (S. Jahn, personal communication, 2007).

Although one can provide innumerable definitions of the landscape, according to the theory and the social component, there is a formula that could summarize all of them:

$$\mathbf{L} = \mathbf{S} + \mathbf{N}$$

In the expression above, there are three essential factors described, of which a landscape (L) is composed of: the subject (S), the nature (N), and the relationship between both (Assunto, 1976). This notion is also treated by the ecology discipline that studies the interrelation between human society and its living space. The landscape, in this ecological point of view, is intended as a set of interconnected ecosystems that externally act with the needs of the society (Naveh & Lieberman, 1994).

However, Malcolm Budd, a philosopher who analyzes the landscape not based on relationships with man (who in the previous case acted as a protagonist in relation to nature), disagrees with this statement that is based on his individual and unique character. He is in favor of a positive aesthetic that accepts all nature as beautiful, since it is not altered by man (Budd, 2002), exclusively accepting its natural changes, due to time, day, and night and the seasons of the year (Krebs, 2014). This is a concept that is definitely related to the aesthetic enjoyment of the landscape, which is able to arouse emotions in the subjects who stop to admire it (Table 2.1) (Griffero, 1996).

2.2.3 *Landscape Legislation and the Question of Identity*

As already mentioned, in attempting to provide a definition of the landscape, various disciplines have been discussed to provide an explanation of it in different fields. As the landscape is a versatile theme, it is too simplistic to define it according to a unique expression. However, the reference points for a sharing definition, at least on the operational level, are the European Landscape Convention (*European Landscape Convention*, 2000) and the Legislative Decree 42/2004 (Italian Legislative Decree, 2004), also known as the Urbani Code, with its amendments in 2006 and 2008 (Leone, 2009).

The European Landscape Convention represents the first international treaty that is exclusively dedicated to the European landscape as a whole. It applies to the entire territory of the Parties: in natural, rural, urban, and peri-urban spaces. Therefore, it recognizes, in equal measure, the landscapes that can be exceptional, the landscapes of everyday life, as well as degraded landscapes. It defines the

Table 2.1 Summary of some existing theories about landscape, used in the text (Katia Talento, 2019)

Reference	Definition	Reference	Definition
Corner (2014)	There is nothing natural about the landscape. It is a product of the imagination, anything that is previously imagined and afterward represented in images. It is the result of an architectural project	Turri (2003)	Social aspect of the landscape: It exists because there are those who look at it, giving it a meaning
Simmel, (2007)	Landscape derives from nature, being a part of nature itself	Tilley (1994)	The landscape is the artificial result of a culture that perpetually redefines its relationship with nature, in which the subject is entirely part of it
Paulhan (2016)	We like landscape because we like nature, and the landscape is the portrait of it	Cosgrove (1998)	The social landscape has two ideal typical figures: The "insider," rooted inhabitant, and the "outsider," disinterested spectator
Amiel (1931) and Soares (1982)	Landscape is a state of the soul	Poli (2002)	In the social landscape, there is the figure of a "care- taker" who takes care of the places by choice
Lettini and Maffei (1999) and Andreotti (1996, 1998)	Inner landscape: Relations that unite places with personality and experience	Clement (2008)	It is important to consider the planet earth as a single garden in which man is the gardener, highlighting the global sense of the landscape
Twigger-Ross & Uzzel, (1996) and Jones (1991, 2003)	Description of cultural landscape as a landscape transformed by man, where exist relations between him and the territory	Jakob (2009)	There is a landscape formula which could summarize all the landscape definitions: $L = S + N$ (landscape = subject + nature)
Antrop (2000)	The landscape is a whole that is more than the sum of the parts, being the necessary synthesis for the true understanding of the whole	Budd (2002)	He is in favor of a positive aesthetic that accepts all nature as beautiful, since it is not altered by man, accepting exclusively its natural changes
Domingues (2001)	Landscape identifies itself with the object of study of geography. It is an element positioning between the natural and human sciences.	Griffero (1996)	The landscape, being an aesthetic component, arouses emotions in the subject that admires it
Assunto (1976, 2006)	The landscape is a form that the environment confers on the territory. It is the country considered from the artistic point of view		

landscape in a perspective of preserving the diversity of the landscape itself and privileging the social aspect of shaped landscape in the past by man and nature. Considering recognition and protection, it underlines its cultural, environmental, social, and historical importance. Finally, it invites states to define safeguard, management, and conservation policies for future generations (*European Landscape Convention*, 2000; Kizos et al., 2018; Miccoli et al., 2014). On the other hand, the Urbani Code represents the main Italian regulatory reference that attributes to the Ministry of Cultural Heritage and Activities the task of protecting, conserving, and enhancing Italy's cultural heritage. It defines the concept of cultural asset, including landscape assets, and it proposes the recognition of the notion of landscape within the cultural heritage (Italian Legislative Decree, 2004). In both cases, the fundamental question is clear, and it represents the landscape concept as a contemporary expression of culture and nature, dynamic entities in constant evolution (Leone, 2009).

Focusing the attention on these regulations, it is possible to understand that there is a great richness and diversity of landscapes. While admitting this variety and observing the landscapes, it becomes obvious that history is embedded in the landscape. It connects the natural conditions with a long and specific human action, which reflects both the natural and cultural history of a territory.

One comes to the question of the spirit of the place, a dynamic character that is constantly changing, but it is unique to each landscape, and it integrates into the local identity (Antrop, 2000; Antrop & Van Eetvelde, 2017; Usher, 1999; Washer, 2003). However, corresponding to a product of the past, the current landscape constitutes a record of the collective memory (Ribeiro, 1998) and a powerful element of cultural identification, such as language and religion (Gaspar, 1993), as well as connection with the territory and with the identity that it represents (Gustavsson & Peterson, 2003; Janin, 1995; Pinto-Correia & Ramos, 2000; O. Ribeiro & Lautensach, 2009).

The identity of a place, as the American psychoanalyst and philosopher James Hillmann suggests, is "*the soul of a place*," a character that belongs to a place and to no one else (Hilman, 2004). It is the product of past events revisited in the light of creative acts, also dictated by social, cultural, and economic changes. In this sense, the concept of identity is very closely connected to that of Marc Augé, according to which a place produces identity when the community, which resides within it, recognizes itself in it (Augè, 1992). The relationship between space and community, which is also contained in the European Landscape Convention, is based on the concept of the "soul of the place." The landscape is similar to a palimpsest and a stratification of signs between the deletions and rewrites (Raffestin, 2012; Shama, 1995). The binomial landscape-identity, as applied to the contemporary era, often leads to talk of crisis of legibility, representativeness, and planning, because the contemporary landscape is perceived or described as "ugly," disharmonic, the same in every place and devoid of identity. Therefore, this element loses the character of uniqueness that previously distinguished it, creating disorientation. It is important to recognize the identity of a place, because it enables contact with the traces of the past of that place, and, at the same time, it is crucial for the prefiguration of the future, for its ability to arouse emotions and with them new identities (Cillo, 2008).

According to Bernard Lassus, who has dedicated an important part of his theoretical and planning work to this topic, the landscape consists of the stratigraphy of several historical chronologically overlapping layers (Lassus, 1998). A kind of "millefeuille" (Venturi Ferraiolo, 2006) to be patiently and precisely dug and reinterpreted, using the tools of poetic archaeology. According to Massimo Venturi Ferraiolo, "*every landscape is archaeological as such*." He also adopts the metaphor, which was designed by environmental and anthropological transformations, while Weilacher highlights how the landscape, over time and maturing, becomes a cultural landscape through the invisible overlap of historical traces (Matteini, 2009; Weilacher, 2005).

> In the flesh of the landscape all the stigmata of the past are impressed and endured. Landscape is a memory and I can interrogate it. (Cillo, 2008) (M. Corajoud in B. Cillo, 2008, p. 87)

2.3 Landscape Mutation—Abandonment and Residual Spaces

2.3.1 Changed Landscape

The concept of safeguarding the identity of a place or a landscape, as it was previously described, is not compatible with our contemporary era, where, on the contrary, one looks at the landscape identity with a nostalgic feeling of an important factor that has been lost. Against this image, the experience shows nature as not only humanized but also effectively and visibly altered, because of constant transformations. The image of a world that is dominated by human presence usually prevails, which is accompanied by the transformative capacities of machines (Veríssimo Serrão, 2011). The growing industrialization has significantly changed the usual territorial structure and, at the same time, provoked the first serious environmental issues. An example is the construction of the railway network that required massive interventions, such as the construction of countless viaducts, bridges, tunnels, embankments, etc., with the aim of leading tourists to places that are cataloged by the touristic system (Bèlanger, 2009; Jakob, 2009).

Industrialization has also generated other types of landscape, having not only a recreational and entertainment purpose but also a productive one. Thus, one witnessed the birth of the so-called industrial landscapes, being dotted with the presence of innumerable warehouses and buildings, intended for production, which modified the original appearance of the territory. Usually, industrial buildings, by their nature, use, shape, and size, are often considered to be extraneous volumes to the landscape, as they are difficult to integrate into the context. Although they are no longer mere production sites and take on a plurality of functions (office, warehouse, and supermarket), they are negatively perceived by the local communities. For this reason, a particular "Nimby" syndrome is discussed, or "not in my backyard" or "not at my house," when planning a new production site, only not to encounter new industrial volumes that do not integrate into the existing landscape (Cipriani, 2012).

In relation to various economic activities, one distinguishes the industries of products that are characterized by the materiality of the asset object of their activity (agricultural, manufacturing, and commercial) and service industries (credit, transport, security, and education). The industries are considered in the strict sense, the extractive (that exploits the mineral riches of the subsoil), the manufacturing or transformation industries (that transform raw materials or semi-finished products into finished products, such as the chemical or mechanical industry, electronics, food, textiles), and finally the energetic (that transforms the energy produced by the different sources). The extractive industries are those that have modified, and still modify, the landscape in a more invasive way. They are industries in which the effects of industrialization are much more visible, because, while the others only cause the addition of new industrial buildings on the territory, the extractive industries alter its morphology instead, deforming it according to the various production needs and "disfiguring it." Therefore, they cause a strong impact on the territory, modifying its physiognomy, but also its social structure, economy, and environmental quality.

Many quarries are subjected to closure or abandonment for various reasons after being abundantly exploited (Figs. 2.5 and 2.6). The most frequent ones are due to the difficulty of selling the extracted material and the economic incapacity that is caused by international competitiveness. Among other causes, one also finds the failure, the depletion of the extracted material, questions of inheritance (in most cases the owner has no descendants), or, to conclude, the presence of magmatic rock inside the exploration area that causes the metamorphose of marble, contaminating it, reducing its quality, and altering, for example, its resistance and color. The crisis in the building sector has also contributed to reducing the quantities of the extracted stone materials.

The abandoned quarries are a type of landscape that can stand as an emblem of the gap. They were first exploited and then discarded, without following any design indication but rather the imperativeness of productivity (Tsolaki-Fiaka et al., 2018). To cause their structure to be even more casual and problematic, there is also a lack of a reflection on the landscape or landscape studies consistent with the era of the birth of places, with the consequent random overlapping of the traces of past uses. The quarries are destined not to be able to return to their previous state, subject to the violent removal of materials and the desire to gain profit. The restoration of the initial conditions is always impossible, leaving the remaining fragments in the territory to be the fundamental elements of the design approach. Considering this reflection, the quarry lends itself to a plurality of interpretations, being, at the same time, a collection of natural and artificial fragments, presence, and absence, firm but constantly evolving – a place that escapes univocal reading – and this is why it leaves room for many futures (Spinarelli, 2016).

Fig. 2.5 Abandoned marble quarry in the Estremoz Anticline, Portugal (Katia Talento, 2018)

2.3.2 Landscapes of Abandonment

2.3.2.1 The Friche Concept

Over the years, many of the countless industrial areas, previously described, have lost their business, turning into brownfields, due to the lack of economic funds or depletion of exploration material. Most often, they are a burden for municipalities

Fig. 2.6 Abandoned marble quarry in the Estremoz Anticline, in Portugal, with the waste heap in the background (Katia Talento, 2018)

that fail to make them profitable, but, at the same time, they are a virtue for those who can see their potential for re-use. These unstructured spaces are identifiable with the industrial term "*friches*." Most countries are still in the process of defining what is meant by "*friche*" (brownfield), but, in general, it is an area previously used for industrial purposes, which is now abandoned and unused. In the past, the term was used to indicate the uncultivated fields that awaited the new plowing and then evolved over time (Chaline, 1999). The term also denotes, up to now, at the juridical level, all the areas where human activity ended, such as the abandoned industrial activities.

Claude Chaline, a French urban planner, identifies different types of *friche* during different historical periods. In the French and Italian circles, the church increased its real estate for a long time because of the benevolence of rich and illustrious individuals and donations of the faithful. This led the church to control a large portion of the territory of the city. At the end of the eighteenth century, the assets of the church were limited according to different methods, even though expropriation, and a large case of urban *friches* was generated. In fact, the buildings often remained unsold or unused for a long time, which created a widespread condition of abandonment. Little by little, the state took charge of these assets by transforming them into public and administrative buildings. A similar phenomenon occurred in the nineteenth century when a vast patrimony of disused military installations was found at the end of a long period of wars and armed conflicts.

In the same way, in the different evolutionary phases of the history of the city, other types of *friche* have been generated, such as those that are related to the world of transport, thus referring to the stations, depots, and railway tracks. In addition to the disused railway goods, one must also consider the places of degradation that the railway lines generate along their margins, often posing as an obstacle. Additionally, the port *friches* derived from the same process, and all those with functions that are incompatible with the city, were trapped during their growth as airports too small for passenger traffic and too close to the dense city. Large buildings for agri-food (slaughterhouses, barns, etc.), department stores, and small shops failed to change the system of sale and deposits used for trade. All of these phenomena have given rise, since the middle of the twentieth century and especially in the last 15 years, to a new meaning of the word that no longer only refers to the decommissioned industrial heritage, but also the areas that have become waste of the city, located inside it (Chaline, 1999; Di Marco, 2016).

It is important to remember that the aforementioned "*friches*" are not only perceived as positive natural elements, but they can be seen, above all, as a depreciation of the landscape, an aesthetic, economic, and ecological loss. This perception often appears when urban planning is wavering and leaving gaps, or even large scars, in the urban fabric (Dumesnil & Ouellet, 2002; Sénécal & Saint-Laurent, 1999, 2000).

2.3.2.2 Wasteland and the Notion of Terrain Vague

If one wants to solely focus on the residual material, one can say that cities are invaded by waste and garbage in all forms and from different points of view (Campos et al., 2013). The wastes of industrial society and, in a special way, those of the consumer civilization are an essential and highly complex component of the life cycle of all material goods. They are in some way the "removed," the "expulsion" of that systematic activity of robbery, and the wastes of the earth's resources on which this type of society is founded (Viale, 2000). In the early 1970s, Kevin Lynch was among the first to address the issue of waste and garbage that are produced by the company in the United States, arguing that the gap and decadence are inevitable consequences (voluntary or otherwise) of the contemporary urban phenomenon and its development. He believes that waste is something useless or useless from the human point of view; it is loss, abandonment, decline, separation, and death (Lynch, 1990). When one talks about scraps, one is led to think and associate them with garbage, waste produced at homes, or at most waste from industries and therefore related to pollution. Even today, the association of waste with the spatial context is not immediate, but one must not forget that there are also places considered as waste and that waste can also take on a spatial, territorial, and landscape dimension (Di Marco, 2014).

Among the waste that is produced by the city, there are entire areas that are expelled because they lack use, definition, and form. They are abandoned or disordered industrial spaces, border areas, and "empty" lands unrelated to the urban system, excluded from it, but physically internal to the city, which present themselves as "absences" in the contemporary urban fabric. They function in various ways as shelters and mirrors that allow us for to examine our daily environment, outside the frenetic circuits of work, commerce, transit, and ourselves. They are also containers of a fragmented shared history, which illuminates the imperfect process of memory, and constantly tries to remember and reconstruct the past (Hall, 2013). Based on the work of lgnasi de Solà Morales (De Solà Morales, 1995), Catalan historian and architect, we can also classify these elements as "*Terrains vagues*," a word that has consistently gained popularity, especially in the English language. The term "*Terrain*" in French indicates a delimited portion of land as a lot or a field, while the term "*Vague*" contains three great connotations: *Vacuus*, meaning empty; *Vagus*, meaning uncertainty; and *Woge*, meaning mobility, fluctuation, and also indicated its state of being in between time and space (Pineiros, 2016).

The complete term, "*Terrain vague*," contains a multitude of meanings, and it is used for indicating various types of wastelands located within the confines of the city. At the same time, it is external to the daily uses of citizens; "*Terrains vague*" seems to be almost forgotten spaces in which few values survive and in which the reminiscences of their past have taken precedence over the present, which makes them almost like ruins and fascinating spaces. Therefore, this name reflects the duality of these landscapes, being essentially something that is not designed but, at the same time, having a high power of transformation that is precisely given by not being constructed; waiting spaces, spaces of the possibility that enclose in itself a transformation in power (Di Marco, 2016). The recognizable "*Terrains vague*" forms are those identified with the former industrial sites, such as abandoned factories and quarries, gas meters, water towers, shipyards and marshaling yards, military areas, mining plants, general markets, warehouses, canals and locks, hospitals, schools, sewage treatment plants, bridges, tunnels, and dry docks (Barr, 1969; Boeri, 1990; Davis, 2002; Secchi, 1989).

Secchi (Secchi, 1989), for example, provides well-known cases of "*Terrains vague*," subsequently redeveloped, ranging from the port areas of San Francisco, Rotterdam, Liverpool, and London to the sprawling Fiat Lingotto factory in Turin (renovated in the late 1980s in a multi-use complex according to a Renzo Piano plan) and the first Modena racing track (now Parco Enzo Ferrari). Duisburg-Nord Park in Germany by the Latz & Parter studio, the New York High Line by architect James Corner, and the Nordbahnhof Park in Germany by the Fugmann & Janotta studio are other examples. These are some references of an emerging group of architectural projects that are related to waste landscapes (wastelands) or abandoned industrial sites (derelict lands), which share the same aesthetic characteristics and the particular quality of creating a balance between the past and the present (Jonas & Rahamann, 2015).

In his essay, Solà-Morales advocates for a much subtle way of integrating the past of abandoned or destroyed places with architecture proposals. In a way, Solà-Morales' architectural approach toward "*Terrain vague*" resonates with the aesthetics of ruins, due to their insistence on maintaining the form of a construction of the past without completely renovating them. This is even more evident in the landscape designs, aforementioned, involving derelict spaces that integrate the industrial ruin (Pineiros, 2016). Concluding, the "*Terrains vagues*" are daily areas that expose the stratification and the palimpsests of the nature of all places. They are what the architect-artist Stalker (Stalker, 1996) defines, in his manifesto, as *"spaces for comparison and contamination between organic and inorganic, between nature and artifice,"* which constitute the negative of the built city, the interstitial and the marginal, spaces abandoned or in the process of transformation. Or they are what Secchi (Secchi, 1993) defines as *"immense collections of objects positioned tactically next to each other, dumb"* (Barron, 2013).

Many other terms have been used to describe abandoned and residual spaces: "Derelict land" (Barr, 1969); "Zero panorama," "Empty or abstract settings," and "Dead spots" (Smithson, 1996b); "Vacant land" (Bowman & Pagano, 2004); "Wasteland" (Gemmell, 1977); "il Vuoto" ("the Void") (Secchi, 1989]); "Urban wilds" e "Urban sinks" (Lynch, 1990); "New, nameless places" (Boeri et al., 1993); "Dross" (Lerup, 1994); "Drosscape" (Berger, 2007); "No-man's land" (Leong, 1998); "Dead zones" and "transgressive zones" (Doron, 2000); "Superfluous landscapes" (Nielson, 2002); "Spaces of uncertainty" (Cupers & Missen, 2002); "Le Tiers-Paysage," "Les delaisses," "the Third Landscape;" and, roughly, "Leftover lands" (Clement, 2003). Other common terms, among others, include "Brownfields," "In-between spaces," "White areas," "Blank areas," "SLOAPs" (Spaces Left Over After Planning), "Voids" (Jonas & Rahamann, 2015), and "Terra incognita" (Bowman & Pagano, 2004). Despite the multiplicity of definitions and terms, these landscapes are not united by physically belonging to the city but from having been expelled from it for the conclusion of their life cycle, for the incompatibility of their use with the urban reality, or for the loss of their economic value. These territories, "rejection of the city," do not have an intrinsic negative value but a specific characteristic, natural consequence of the process of growth and development of the city, and its metabolism (Di Marco, 2016).

2.3.2.3 Drosscape as a Refuse Space

The waste can be considered as a natural component of an evolving and dynamic city, and, as such, it is an indicator of its healthy growth (Berger, 2007), metabolism (Gandy, 2003), and part of its life cycle. Thus, the control of urban metabolism and the production and management of waste are fundamental aspects to be taken into consideration in the growth and development of a city. The waste landscapes are the places that are rejected by society due to the presence of functions that are not compatible with it: abandoned areas, planning residues, and places where productive processes have been exhausted by dissipating, consuming, and losing identity. The common derogatory and critical interpretation of these places emphasizes the loss of their environmental qualities and places them in strong contrast to the naturalistically higher-quality landscapes.

For this reason, these places require greater attention from the communities and institutions for the realization of the regeneration project (Berger, 2002). Working with these territories of waste and considering them as part of the landscape means working on the most fragile and problematic part of the city, triggering a design discourse on the landscape and the recovery of urban metabolism products through their treatment. The energy needed for urban regeneration must come from the territories rejected by the city because the future of growth and transformation of cities is closely linked to the reuse and re-imagination of waste landscapes. In this context, the designer's role is to integrate this inevitable waste into a more flexible and aesthetic strategy, creating a complex, integrated, and dynamic plan (Di Marco, 2016).

Although many urban planners have been interested in urban waste sites and their definition, a detailed classification of the different phenomena is still missing in the European landscape. Alan Berger, who is an American urban planner, was the first to attempt a rigorous cataloging of the different places of waste, in the variegated American territory, defining them as "*drosscapes*" (Table 2.2), identifying different categories that generated them:

I. **Wastelandscape of dwelling** (LOD): it refers to voids of land that are integrally designed into housing developments, especially into walled or gated enclaves. These voids often have singular programmatic intentions (golf course, buffer zone, preservation area, trail system, etc.). There are two types of LOD voids: those positioned outside and inside the enclave. "Outside" voids encircle the enclave as buffers and separators from adjacent development or other possible nuisance land uses that may affect the quality of life held by the dwellers of the enclave. "Inside" voids are designed, for example, to allow for public utility easements that cross throughout enclave territory. They serve the social, circulation, and recreation needs of their inhabitants.
II. **Wastelandscape of transition** (LOT): it reveals the transitory nature of capital investment and real estate speculation. Some LOTs are intentionally designed and built as transitional land uses, such as staging areas, storage yards, parking surfaces, transfer stations, etc.

Table 2.2 Summary of the terms used to describe abandoned and residual spaces (Katia Talento, 2019)

Reference	Definition	Reference	Definition
Chaline (1999)	Friche	Lerup (1994)	Dross
De Solà Morales, (1995)	Terrain vague	Leong (1998)	No man's land
Berger (2007)	Drosscape	Doron (2000)	Dead zones and transgressive zones
Barr (1969) and Kivell & Hatfield, (1998)	Derelict land	Nielson (2002)	Superfluous landscapes
Smithson, (1996a)	Zero panorama, empty or abstract settings, and dead spots	Cupers and Missen (2002)	Spaces of uncertainty
Bowman and Pagano (2004) and Northam, (1971)	Vacant land	Clement (2003)	Le tiers-Paysage, les delaisses, the third landscape, and leftover lands
Gemmell (1977) and Nabarro & Richards, (1980)	Wasteland	Jonas and Rahamann (2015)	Brownfields, in-between spaces, white areas, blank areas, SLOAPs
Secchi (1989)	Il vuoto (the void)		Voids
Lynch (1990)	Urban wilds and urban sinks	Bowman and Pagano (2004)	Terra incognita (unknown land)
Boeri et al. (1993)	New nameless places	Cupers and Missen (2002)	Spaces of uncertainty

III. **Wastelandscape of infrastructure** (LIN): it includes the landscape surfaces that are associated with the infrastructure, containing easements, setbacks, and rights-of-way associated with transportation (such as highway corridors and interchanges), electric transmission, oil and gas pipelines, waterways, and railways.

IV. **Wastelandscape of obsolescence** (LOO): it refers to the sites that are designed for accommodating consumer wastes. These include municipal solid waste landfills, wastewater-treatment facilities, and "car dismantlers."

V. **Wastelandscape of exchange** (LEX): it describes shopping centers and all those urban complexes where the commercial, but also catering, fitness, and entertainment, functions are concentrated. They are boxes that are surrounded by parking lots and can only be reached via expressways. They generate many interstices and waste spaces, and often they become places of waste when they become unsuccessful and lose their economic value.

VI. **Wastelandscape of contamination** (LOCO): it includes public and federal installations, such as airports, military bases, ammunition depots, and training grounds, and sites used for mining, petroleum, and chemical operation (Berger, 2007; Rathje & Murphy, 2001).

These accurate classifications suggest that it is necessary to reprogram the drosscapes with social plans to transform waste into more productive urbanized landscapes to some degree. They demand a phased implementation of the design that offers "clean" or "green" types of urbanization because they are not immediately entirely occupied (Alker, 2000; Corner, 2004, 2006). In this regard, it is possible to consider and affirm that the concept of the territory is manipulated and modified by man replaces nature, being explored, and conquered over the centuries. On one hand, destruction (the territorial wound inflicted on nature by industrialization) generates a radical dualism with the good landscape, grown over centuries, but, on the other hand, it generates the new non-places that are produced by mechanical civilization, which are places aesthetically ignored and removed. Summarizing, the new interest in the landscape once again passes through loss or crisis, because it is precisely through the latter that one understands the importance of the natural element and one feels the need to act in order to restore the lost places (Jakob, 2009).

2.4 Re-designing the Degraded Landscape

2.4.1 Recycle and Reutilization of the Waste Material

After giving a definition of the abandonment landscape, while using distinct terminologies for the description of this type of scenario, one can state that the territories need to be reactivated and re-introduced in a cycle in which the waste that they produce is no longer a problem to be solved but an opportunity to restart, regenerate, and transform a territory. This would also allow creating new economies that would predict benefits, not only quantitative but also qualitative, for spaces and society.

The transformation derives from the urgent demand, coming in turn from contemporary society, to find ways and methods to stop the phenomenon of land consumption and waste of resources. This action is activated through an "eco-logic" that was inspired by the concepts of the Reduce-Reuse-Recycle triad, now widely established in the field of the so-called Green Economy. In the strategies of urban and landscape regeneration, to the three "R" of the so-called "eco-efficient recycling," the three "E" can be combined: Economy, Equity, and Environment – in other words, economic growth, combined with social equity and respect and protection of the environment (Marini & Santangelo, 2013).

This is how the recycling project was born, which extends the notion of regeneration, and it defines a new context of reference, being linked to the values and materials of ecology, the environment, and the landscape. Recycling means putting back into circulation, reusing waste materials and items that have lost value or meaning, and promoting the concept of sustainability. It is a practice that allows for reducing waste, to limit the presence of waste, to reduce the costs of its disposal, and to contain the prices of the production of new interventions. In other words, to recycle means to create new value and new sense, another cycle, and another life with sustainable solutions. This is where the propulsive content of recycling resides: an ecological action that pushes the existing into the future, transforming waste into prominent figures (Marini & Santangelo, 2013).

In the cultural debate, the theme of the reuse or recycling of abandoned spaces in the city is increasingly proposed and is addressed as an opportunity in terms of environmental sustainability. At the Venice Biennale in 2008, a well-known American economist Jeremy Rifkin already presented his Charter for the Architecture of the New Millennium, in which the recycling of old spaces is assumed as a pillar of the Third Industrial Revolution. The reference is a model of a city that can recover material from existing values and the built environment, especially in the presence of brownfield sites, which eliminates waste processes as much as possible. The challenge is to be able to trace the ability to make new life cycles germinate in the "dead nature" of our increasingly cemented territories in architecture, urbanism, and landscape. It is curious to note that what one calls "*natura morta*" in Italian (dead nature in English, and even earlier in Dutch with the word "*Stillleben*") represents "*still life*" or is "*quietly alive*." This could be the symbol of the present research: transforming inert matter into a resource for new life cycles, to the way in which Morandi transfigured his "dead nature." This is the profound meaning of the concept of recycling (Marini & Santangelo, 2013). As Pippo Ciorra wrote in the introduction to his exhibition at MAXXI Museum:

> Rebuilding instead of building: building on, around, inside, on, with waste materials; to live in ruins instead of building; re-naturalize rather than re-urbanize. (Marini & Santangelo, 2013) (P. Ciorra in S. Marini, V. Santangelo, 2013, p. 13)

The topic is obviously not new, but the concept of recycling as applied to the themes of city and landscape architecture can go from a purely technical term to a keyword to seek for renewed strategies and design tools for the regeneration to which one aspires. It considers not only the waste materials of the recent transformation processes but also the same "inert" fragments of pre-existing territorial geographies, which are involved in processes of abandonment, marginalization, and rejection. The historian Antonella Tarpino has rightly called these territories "disoriented," that is, fragile territories or traces left as an inheritance (Tarpino, 2012). In this vast field of reflection, it is important to emphasize the recognition of the landscape in every place as an important element for the quality of life of the populations: *"in degraded areas as in those of great quality, in places considered as exceptional, as in those of life daily,"* as stated in the preamble to the European Landscape Convention.

The goal is to use these "artifacts," material or immaterial, for the purpose of re-reading landscapes that provide meaning and identity to these territories, even in their new touristic dimension. This considers "system projects," rather than individual interventions, for the reuse or recovery of specific abandoned areas or factories (Corner, 1999, 2006; Fabian, 2012; Marini & Santangelo, 2013; Waldheim, 2016). Discussing the city as something that can be recycled leads to considering its rhythm cycles of life and metamorphosis. The famous words of Jane Jacobs should be interpreted in this sense as, according to her, the city does not follow an unmodifiable biological path, but it has the ability to regenerate within itself to overcome a cycle of life and decline, reinterpreting and constantly renewing itself (Jakobs, 2002; McDonough & Braungart, 2002).

2.4.2 Rehabilitation Concept—A Bridge Between Past and Present

The words described above belong to an optimistic attitude that prefers rewriting the meaning of the waste concept using material recycling operations. In this way, the violated landscape acquires a greater importance, contrary to what happened in the previous years, where the attention has always been paid to the natural, original, gorgeous, and untouched landscape. The "*Decoupage ferroviaire*" of Cezanne (1870–1871) canvas, showing the wound inflicted on the landscape by the construction of the tracks, remains, from this point of view, an extraordinary document, in the strongest meaning of the term (Fig. 2.7). A real awareness of the destroyed and polluted earth that will be later analyzed in some films and in the land art of the 1960s.

The works of Michelangelo Antonioni are the first in the history of cinema to systematically give the landscape the role of protagonist. His films cover the totality of the sublime and violent nature, until considering the heavily industrialized territories of the post-war period. The Ferrarese filmmaker, fascinated by the unusual aesthetics of industrial sites and landfills of all kinds, pays attention to what was almost always excluded and visually censored, that is, to the other nature, modified, cut out, and recomposed to make it unrecognizable. Antonioni will allude to the tradition of the postcard landscape, to the perennial and petrified image of nature for tourism use, and to the irruption of the landscape that is experienced as a surprise (Jakob, 2009). Having these images as a reference, it is possible to say that natural beauty is now mutilated, and its landscapes are degraded (Grimaldi, 2011). One has

Fig. 2.7 Paul Cezanne—Decoupage ferroviaire (1870–1871) (Wikipedia, 2019)

moved from a "figurative city," which is well organized by design rules and architectural models, to a "disfigured" city, characterized by invisible and abandoned segments, excluded from the infrastructure network (Boyer, 1995). Summarizing, according to the words of French landscape architect, René-Louis de Girardin, one can say that the beautiful nature is threatened, and one must protect it from humans (Pierre, 1982).

In this sense, the intervention of landscape architect is strictly necessary toward intense humanization, which becomes quite inhumane (Cabral, 1993). According to Corajoud, a French landscape designer, this human intervention should not consider the landscape as a *"tabula rasa"* and, consequently, seeing s it as a simple "blank page" where we can unilaterally print our ideas. On the contrary, it is necessary to consider it as a place that already has a history and characteristics that the landscape architecture project should regulate. The project is guided by the landscape itself, and the work emerges as a process of co-initiation between the creativity of the existing subject and object (Veríssimo Serrão, 2011). In relation to the issue of pre-existence, there are some authors interested in the topic. Laurie Olin states:

> The only thing introduction that we can ever know for certain about the world is that which exists now or has existed in the past. To make something new we must start with what is or has been and change it in some way to make it fresh. How to make old things new, how to see something common and banal in a new and fresh way is the central problem. (Olin, 1988)

As Corner says, "*the past is like a repository of meaning that may be mined to prompt new possibilities that transcend the given present for the construction of new horizons*." To the author, landscape is something that is consistently made and remade in the perpetual process of transformation (Corner, 1999).

Rehabilitating, or rather, re-finding a place means projecting the landscape in the past and in the future because it cannot subsist without transmission of knowledge, culture, and tradition (Bonesio, 1999, 2009). The tradition does not refer to some vague recollection of the past, frozen, and inaccessible but to the creative and processual power of which we are an integral part. Consequently, tradition is a dynamic artifact, being a result of human work and the accumulation of ideas. Tradition forms the basis for any future work. The objective is to devise new meanings (futures) from a critical and yet imaginative reinterpretation of our tradition (past). Much of the difficulty in contemporary landscape architecture lies in such recoding and transformation. How do seemingly disparate and banal meanings in the landscape discover new life and usage, renewing an art form while also maintaining its tradition? In other words, how can one turn the ordinariness of everyday situations into something imaginative or fresh, pertinent to our time but not estranged from tradition?

The landscape itself is a text that is open to interpretation and transformation. It is distinguished from wilderness, which is land that humans have modified. Moreover, landscape is not only a physical phenomenon, but it is also a cultural schema, being a conceptual filter through which our relationships to wilderness and nature can be understood. It is a phenomenon beyond immediate comprehension

(Veríssimo Serrão, 2011). While recognizing its remarkable value, landscape architecture has always had a fundamental mission of restoring the balance of "living works" on the dead ones. It will have to adapt rather than replace and make the most of what nature has given us; just as the old industries will have to be integrated into the general landscape, not forgetting the presence of the past that offers a sense of completion, of stability, and permanence, in resistance to the rapid pace of contemporary life (Bonesio, 1999, 2009; Cabral, 1993; Lowenthal, 2015; Olin, 1988).

In this sense, it is better not to build new landscapes but to recover the forms of territory prior to modernist and industrialist barbarism (Cervellati, 2000) and providing a new face to the factor of abandonment, as also stated by Joachim Mitchell. His work explores the possibility that new buildings arise from the reuse of rubble and waste, which translates the problem into a resource (Mitchell, 2010). The landscape is full of resources; it is inexhaustible in the sense that it offers a multitude of clues that tell us what it is, what it was, and what it can become. Indeed, in the flesh of the landscape, all the stigmas of the past are imprinted and endured. The landscape is a memory, and one can question it (Corajoud, 2009). While taking into consideration the words aforementioned, it is possible to understand that the relevant element of a landscape is its context. In fact, when a word or phrase is taken out of context, it can become meaningless or its meaning can become distorted, even to the extent of implying the opposite of what was originally intended. As Tim Waterman, landscape architect and theorist, said:

> When we live in a place, make a home in it, a permanent investment, we are said to inhabit it. A good place is one in which we feel comfortable, that fits us like a pair of worn jeans. (Waterman, 2009)

Based on his statement, it is evident that context is not less important in the landscape. The landscape provides context for everything that is built and for the activities of our daily lives. Anything that is built in the landscape needs to consider its surroundings and situation to be successful and sustainable, and this work is at the core of the practice of landscape architecture. Everything in the landscape is part of an interconnected system that forms the fabric of our existence. If any element in the landscape is altered, everything else is also affected. Accordingly, the landscape can no longer be solely considered as decoration around the base of buildings (Waterman, 2009).

2.4.3 The Promotion of Art and Landscape Architecture in the Abandonment Territory

The question of the context, a fundamental element of a landscape, gave much more visibility to the landscape architecture that has experienced enormous growth and gained visibility in recent years (Waldheim, 2016). It occurred due to the design of new parks, waterfronts, squares, public places, gardens, and the revitalization of many urban spaces around the world. It has also transformed into a therapeutic tool that has repaired the innumerable damage in the territory, in abandoned places with undefined use (old warehouses, factories, landfills, etc.), and invented new spaces that are more suited to life in the name of industrial decline (Meyer, 2017). There are several other factors that have recently promoted landscape and are associated with subjects of site, environment, and new technologies. Landscape is increasingly sought for its unique and intrinsic characteristics – its scenery, history, and ecology. Whether as a theme park, wilderness area, or scenic drive, the landscape has become a huge, exotic attraction by itself, a place of entertainment, fantasy, escape, and refuge.

In the 1970s, these abandoned areas never healed mainly brought to the center of the debate with the contribution of artists who find in these places an island within the city where they can freely express themselves. Many photographers, like John Davies e Manolo Laguillo, use these places as subjects of their works. The artistic movement is in favor of the conservation of these territories, to give space to something different, of unexpected, of not accepted by the canons of the city (Di Marco, 2016). In this context and with the following line of thought, one also witnesses the emergence of land art. In the hands of artists, such as Smithson, Heizer, Walter De Maria, Christo, Robert Morris, Herbert Bayer, and James Turrell, landscape is less a scene for contemplation (Veríssimo Serrão, 2011). They have given a completely new meaning to the concept of nature, realizing their installations in industrialized or heavily polluted places to discover the surprising attraction of uninhabited and dangerous places (Fig. 2.8). Their paradoxical works do not hesitate to transform nature with the use of heavy vehicles previously managed by industry. The artists agree to be on the side of the artifice, indulging in the presence of the "machine in the garden" of civilization, which amplifies it further. Land art refers to the landscape through a constant strategy of displacement and disorientation. Its most important effect consists of making every representation impossible, disassembling the image. In this way, the works are exposed to natural elements, and they are no longer rendered eternal within museums, denouncing any attempt at documentation. Therefore, the piece only exists as a perpetual transformation (Smithson, 1979).

This concept has allowed for a radical redefinition of the landscape, also including the word "gap," which has stopped assuming a negative meaning. As Bernard Lassus argues:

> Today one can very easily imagine that a polluted place produces a beautiful landscape and that in the opposite a non-polluted place is not necessarily beautiful. (Lassus, 1991)

The idea of the unchangeable and insensitive landscape to change is lost. This situation clearly opens the possibility of the landscape to absorb the time dynamics, by elaborating strategies that are capable of dialoging with the wastes, rather than excluding them a priori as an undesired subject. In addition, Gilles Clement, a French landscape architect, sees the residues as part of a dynamic process without interruption and in which the *"appeal to architecture still seems the only appropriate way to affect the natural disorder* " (Clement, 2003).

Fig. 2.8 Michael Heizer. Dissipate. 1968. Nevada Desert. (Wikipedia, 2019)

This thought establishes new operational scenarios and lands that existed already, but our gaze was not used to perceive them. The future of these white areas, always waiting for a precise role, can therefore be to form areas of change, which are ready to be activated in the case of necessary revisions of the existing structure (Marini, 2008a, 2008b, 2010). The landscape is transformed into a palimpsest for future scripts that overlay, and so the project becomes the tool that interacts and creates relationships. The reading of the waste and discarded landscapes is essential to act on them (Kenneth, 2010).

With the experiences of photographers, it has been possible to ascertain what enables these places to go beyond the traditional aesthetic categories, upsetting our way of looking, recognizing, and qualifying. An example is Bernd and Hilla Becker with the typological cataloging of industrial artifacts, or with photographs by Naoya Hatekeyama, which highlight the signs left in the terraced quarries that seem to be negative cities. Therefore, photography is an instrument of knowledge, and it becomes fundamental in giving an unusual reading of the discarded spaces, which will, in turn, be of fundamental importance in the intervention process on the waste (Spinarelli, 2016).

Architecture and urban planning are directly involved in the production of intervention tools for the transformation of this waste. It is where their contribution to the conception of new life cycles in architecture, cities, and landscapes becomes essential. Based on this logic of thought, many references to an emerging group of architectural projects, which are related to wastelands or derelict lands, exist (as previously mentioned). As comprehensively discussed in the preceding parts of this research, the current trend in the architecture and landscape fields is related to the revitalization of the industrial abandoned sites through the reuse of the residual matter. The purpose is to create "new" territories promoting the tourism and economy of the countries. Especially, the extractive industry (case study of the present work) is the protagonist of this transformation, which includes different requalification projects with different applications. The abandoned pits can be reused through many ways, such as agricultural applications, resource storage, underground warehouses, and tourism development.

For example, the abandoned pits in Heerlen, Limburg Province, Netherlands, were turned into power stations for utilizing geothermal resources. The abandoned tunnels of Wieliczka Salt Mine, which were located in the suburb of Krakow City in Poland, were rebuilt into an ancient salt mine museum for exhibiting the relics of the salt industry and salt carving works, as well as in the treatment of respiratory diseases. An abandoned gold mine in Blake, South Dakota, America, was turned into a deep underground laboratory to create a deep underground environment for the experiments in the frontier of particle physics. The Butchart Gardens, near Victoria City, Canada, which used to be a limestone quarry, is now the most famous garden in North America. The Stearns Quarry Park located on the South Side of Chicago is an urban leisure park that is reclaimed from an abandoned quarry and a construction waste landfill site. The Garden of Eden in England, which was built on a large, abandoned clay pit, is the world's biggest greenhouse exhibition center of plants. The Intercontinental Shanghai Wonderland Hotel in China is the first five-star hotel that was built in an abandoned pit, and the Ice World and Water Park, IWWP in China, is the largest indoor snow park in the world, which was built in the Dawangshan Pit (Tan et al., 2019). In the Salento peninsula, the Fossil Park near Cutrofiano village was created inside an abandoned clay quarry (Margiotta & Sansò, 2017). In Kosovo, in the municipality of Hani Elezit, 12 new dwellings and green zones were built in the Dimce Quarry after the backfilling and levelling of the depleted closed pit area (Berisha et al., 2017). The sand and gravel quarry in the City of Crystal Lake, Illinois, was reclaimed to host the city-owned Three Oaks Recreational Area, a 66-ha public recreation area, which is popular for swimming and fishing, and, finally, in South Africa, the case study that is probably the best known is the residential and recreational marina development in the reclaimed stone quarry at the Cape Town Waterfront (Vosloo, 2018).

According to the concept of waste reuse from quarries, there are many publications concerned with this problem, which propose ideas for the recycling of the residual material. Its applications are so many, spanning from the glass, chemical, pharmaceutical, environmental, agricultural, plastic, steel, metal, textile, and construction industries to public works. The few applications that can lead to large consumption of this waste material are those that are oriented to civil construction industries, for the production of lime, cement, and concrete (Deepankar, 2018), for calcining acid soils in the agriculture sector, and for the composition of bituminous mixtures useful to road constructions and aerodromes in the civil engineering sector (Deepankar, 2018; Simão et al., 2016; Ulubeyli & Artir, 2015). Another idea is to take advantage of these wastes during the work period when the quarry is active to avoid having artificial unused marble mountains, which cause a strong impact on the landscape. This concept recommends to use, reuse, or discard the marble leftovers for the creation of a new construction material, denominated "Marcrete" (a fusion of marble and concrete), for applications in the architecture's domains (Silva & Esteves, 2015). In summary, all the applications mentioned above treat the marble wastes in the way that they are partially removed from the original place and used elsewhere.

On the contrary, the current work suggests to exclusively act in situ, not exporting any type of material. The case study concerning the Anticlinal of Estremoz in Portugal, an area of 280 km^2 that is composed of about 185 marble quarries, each of which is flanked by a huge pile of wastes (about 15 m high), which are named "*escombreiras*." These giant mounds of remains represent a diverse view to requalify the devastated landscape e to reuse the rests in situ; in fact, they are the main elements to promote tourism in the place through the insertion of new buildings inside them (Talento et al., 2018). The developed interdisciplinary research is based on the concept of giving a new meaning to what is ordinary, working with the waste rather than excluding it, a priori, to preserve the historical and industrial memory of the place, not removing marble residues from the context where it belongs.

In accordance with these theories, is it possible to define, with a single explanation, the landscape? As discussed in the previous sections of the current chapter, it is too simplistic to define it according to a unique expression, but, after these considerations, the landscape can be considered as an alive and ever-changing entity and an infinite sum of individual actions that interpret and modify a place. It is born, evolves, and dies through periods of happy growth, long periods of stasis, and sudden crises, and the community that lives in it recognized herself as in a text of perennial evolution (Zagari, 2006).

2.5 Chapter Resume

This chapter has highlighted the importance of the physical element "landscape" in promoting greater attention toward it. It took several centuries for nature to be represented as an object of pleasure worthy of attention. In fact, the meaning of the word "landscape" has changed over the years, from peripheral, exclusively considered as a mere background in pictorial representations, through central, also gradually acquiring importance in the field of architecture and art, particularly with the birth of land art. This new movement directly acts on the landscape, permanently or temporarily modifying it, with the introduction of enormous installations. These pieces are the result of creative and artistic choices that place the real landscape as the main subject.

The present chapter, while introducing the various disciplines that are related to the landscape, has provided brief but concise theories in this regard to demonstrate the versatility of the natural element, belonging to various fields of intervention. It started by addressing the issue from a pictorial and historical point of view and then proceeded to analyze the main regulations that concern this sector. Subsequently, it has discussed the imaginary component of the landscape; its relations with nature; its geographical, cultural, interior, social, and global identity; as well as its aesthetic aspects. This summary has served to frame the concept of landscape and to address, with solid foundations, the central question concerning the relations between the pure landscape and the disruptive industrial modernity.

Our era includes a different view of the natural territory showing nature strongly humanized, due to the birth of industries and the presence of industrial artifacts located on the land, but also strongly altered in all its appearance. One has witnessed the birth of the concepts of "*Drosscape*," "*Friche*," and "*Terrain Vague*," their scenarios made of machines, wastes, and abandoned landscapes after being widely exploited. They represent a burden for the municipalities that fail to make them profitable but at the same time symbolize a virtue for all those who can see their potential for reuse. The present chapter of the research has focused greater attention on the second aspect having, as its central theme, the redevelopment of these devastated territories through the reuse or recycling of residual matter. The issue has been addressed by analyzing the meaning of the word "reuse," which is important in recovering the existing values of an environment and eliminating waste processes in this way. Several cases of study have been described concerning the requalification of the extraction industrial sites, to make understand the current trend in this field. Some existing methodologies to reuse the residual matter, from industrial activity, have also been illustrated. All these applications treat the marble wastes in the way that they are totally or partially removed from the original place and used elsewhere.

As it will be described in Chap. 5, it is possible to act in a different way. This innovation consists in working with the waste, in situ, rather than excluding it a priori, to preserve the historical and industrial memory of the place, not removing the marble residues from the context where they belong. In Chap. 5, a case study in Portugal will be presented in detail where a new architectural configuration will be conferred to the huge waste heaps located in this area, respecting the surrounding landscape.

Bibliography

Alker, S. (2000). The definition of brownfield. *Journal of Environmental Planning and Management, 43(1)*, 49–69. https://doi.org/10.1080/09640560010766

Amiel, H. F. (1931). *Fragments d'un journal intime* (nouvelle éd). Stock.

Andreotti, G. (1996). *Paesaggi Culturali. Teoria e casi di Studio*. Unicopli.

Andreotti, G. (1998). *Alle origini del paesaggio culturale. Aspetti di filologia e genealogia del paesaggio*. Unicopli.

Antrop, M. (2000). Where are the Genii Loci? landscape—Our home. In *Essays on the culture of the European landscape as a task* (pp. 29–34). Freies Geistesleben.

Antrop, M., & Van Eetvelde, V. (2017). *Landscape perspectives. The holistic nature of landscape*. Springer.

Assunto, R. (1976). Paesaggio, Ambiente, Territorio. Un tentativo di precisazione concettuale. *Bollettino Del Centro Internazionale Di Studi Di Architettura Andrea Palladio, XVIII, 45–48.*

Assunto, R. (2006). *Il paesaggio e l'estetica*. Novecento.

Augè, M. (1992). *Non-lieux. Introduction à une anthropologie de la surmodernité*. Le Seuil.

Barr, J. (1969). *Derelict Britain*. Penguin Books.

Barron, P. (2013). *Terrain Vague: Interstices at the Edge of the Pale*. Routledge.

Bèlanger, P. (2009). Landscape as infrastructure. *Landscape Journal, 28*, 179–195.

Berger, A. (2002). *Reclaming the American West*. Princenton Architectural Press.

Berger, A. (2007). *Drosscape. Wasting land in Urban America*. Princeton Architectural Press.

Berisha, H., Bulica, J., Gashi, S., Kouklidis, K., Mastoris, J., & Dragasakis, K. (2017). *Case study on quarry rehabilitation and land resettlement in Dimce Quarry*. Society for Mining, Metallurgy & Exploration.

Boeri, S. (1990). The narrowing of land uses. *Rassegna, 42*(2), 6–8.

Boeri, S., Lanzani, A., & Marini, E. (1993). Nuovi spazi senza nome (New nameless spaces). *Casabella, 57* (597–598), 74–76, 123–124.

Bonesio, L. (1999). *Geofilosofia del paesaggio*. Mimemis.

Bonesio, L. (2009). *Oltre il paesaggio. I luoghi tra estetica e geofilosofia*. Arianna Editrice.

Bowman, A., & Pagano, M. (2004). *Terra incognita: Vacant land and urban strategies*. Georgetown University Press.

Boyer, M. C. (1995). The great frame-up: Fantastic appearances. In *Contemporary spatial politics, spatial practices*. Sage.

Budd, M. (2002). *The aesthetic appreciation of nature: Essays on the aesthetics of nature*. Clarendon Press.

Cabral, F. C. (1993). *Fundamentos da Arquitetura Paisagista*. Instituto da Conservação da Natureza.

Cagnato, A. (2016, May). Le origini del Paesaggio. *Labsus Papers*.

Campos, M. J. Z., Zapata, M. J., & Hall, M. C. (2013). *Organising waste in the city*. Policy Press.

Cervellati, P. L. (2000). *L'arte di curare la città*. Il mulino.

Chaline, C. (1999). *La régénération urbaine*. Presses Universitaires de France.

Cillo, B. (2008). *Nuovi orizzonti del Paesaggio*. Alinea.

Cipriani, L. (2012). *Vita e morte del paesaggio industriale*. Università degli Studi di Trento.

Clement, G. (2003). *Le manifeste du Tiers-Paysage*. Editions Sujet/Objet.

Clement, G. (2008). *Il Giardiniere Planetario*. 22 Publishing.

Corajoud, M. (2009). *Le paysage c'est l'endroit ou le ciel et la terre se touchent*. Actes Sud.

Corner, J. (1999). Recovering landscape: Essays in contemporary landscape architecture. *Garden History, 28*(1). https://doi.org/10.2307/1587130

Corner, J. (2004). Not unlike life itself: Landscape strategy now. *Harvard Design Magazine, 21*, 32–34.

Corner, J. (2006). Terra Fluxus. In *The landscape urban reader* (pp. 21–32). Princeton Architectural Press.

Council of Europe (2000) European Landscape Convention, Florence 20. 10. 2000. European Treaty Series, No.176. 7 p.
Corner, J. (2014). *The landscape imagination: Collected essays of James Corner 1990–2010*. Princeton Architectural Press.
Cosgrove, D. (1998). *Social formation and symbolic landscape*. University of Wisconsin Press.
Cricco, G., & Di Teodoro, F. (2017). *Itinerario nell'arte* (Vol. 3, 4th ed.). Zanichelli.
Cupers, K., & Missen, M. (2002). *Spaces of uncertainty*. Verlag Muller and Busmann.
Davis, M. (2002). *Dead cities*. The New Press.
De Solà Morales, I. (1995). Terrain Vague. In *Anytime* (pp. 118–123). MIT Press.
Deepankar, K. A. (2018). Feasibility of waste marble powder in concrete as partial substitution of cement and sand amalgam for sustainable growth. *Journal of Building Engineering, 15*, 236–242. https://doi.org/10.1016/j.jobe.2017.11.024
Di Marco, C. (2014, May). Drosscape, un concetto trasmigrante che identifica paesaggi plurali. *Atti Della XVII Conferenza Nazionale SIU – Società Italiana Degli Urbanisti, L'urbanistica Italiana Nel Mondo*.
Di Marco, C. (2016). *La riconversione dei paesaggi di scarto: Saint-Étienne, da cittá nera a cittá del design*. Dipartimento di Architettura, Universitá degli studi Federico II.
Domingues, A. (2001). A paisagem revisitada. *Finisterra Journal, XXXVI, 72*.
Doron, G. (2000). The dead zone and the architecture of transgression. *City: Analysis of Urban Trends. Culture, Theory, Policy, Action, 4*(2), 247–263.
Dumesnil, F., & Ouellet, C. (2002). La réhabilitation des friches industrielles: Un pas vers la ville viable? *Vertigo La Revue Électronique En Sciences de l'environnement, 3*(2) http://vertigo.revues.org/3813
Fabian, L. (2012). *Recycling city. Lifecycle, embodied energy, inclusion*. Giavedoni editore.
Fedeli, P. (1990). *La natura violata. Ecologia e mondo romano*. Sellerio.
Gandy, M. (2003). *Concrete and clay: Reworking nature in New York City*. MIT Press.
Gaspar, J. (1993). *As regiões portuguesas*. Direcção Geral do Desenvolvimento Regional.
Gemmell, R. (1977). *Colonization of industrial wasteland*. Edward Arnold.
Griffero, T. (1996). *L'estetica di Schelling*. Laterza.
Grimaldi, N. (2011). A Estetica da bela natureza. In *Filosofia da Paisagem, Uma Antologia*. Centro de Filosofia da Universidade de Lisboa.
Gustavsson, R., & Peterson, A. (2003). Authenticity in landscape conservation and management—The importance of the local context. In *Landscape Interfaces* (pp. 319–356). Kluwer Academic Publishers.
Hall, M. C. (2013). The ecological and environmental significance of urban wasteland and drosscapes. In *Organising waste in the city: International perspectives on narratives and practices*. Policy Press.
Hanley, D. (2013). *Tacuinum Sanitatis. Health and well being: A primitive medieval guide*. Cognoscenti Books.
Hilman, J. (2004). *L'anima dei luoghi*. Rizzoli.
Ingersoll, R. (2006). *Sprawltown: Looking for the city on its edges*. Princeton Architectural Press.
Italian Legislative Decree, no. 42 (2004).
Jahn, S. (2007). *Cit in national geographic* [Personal communication].
Jakob, M. (2009). *Il paesaggio*. Il mulino.
Jakobs, J. (2002). *The death and life of the great American City*. Random House.
Janin, C. (1995). Peut-on faire l'économie du paysage pour gérer le territoire? *L'agriculture Dans Le Paysage, Une Autre Maniére de Faire Du Dèveloppement Local, Dossier de La Revue de Géographie Alpine, 15*, 11–30.
Jellicoe, G., & Jellicoe, S. (1975). *The landscape of man*. Thames and Hudson.
Jonas, M., & Rahamann, H. (2015). *Tokyo void: Possibilities in absence*. Jovis Edition.
Jones, M. (1991). The elusive reality of landscape. Concepts and approaches in landscape research. *Norsk Geografisk Tidsskrift, 45*, 229–244.

Jones, M. (2003). The concept of cultural landscape: Discourse and narratives. *Landscape Interfaces*, 21–51. https://doi.org/10.1007/978-94-017-0189-1_3
Kenneth, C. (2010). *L'art du paysage*. Arlèa.
Kivell, P., & Hatfield, S. (1998). Derelict land—Some positive perspectives. In *Environment, planning and land use* (pp. 118–129). Ashgate.
Kizos, T., Plieninger, T., Iosifides, T., Martín, M. G., Girod, G., Karro, K., Palang, H., Printsmann, A., Shaw, B., Nagy, J., & Budniok, M. A. (2018). Responding to landscape change: Stakeholder participation and social capital in five european landscapes. *Land MDPI, 7, 14*. https://doi.org/10.3390/land7010014
Krebs, A. (2014). Why landscape beauty matters. *Land, MDPI, 3*, 1251–1269. https://doi.org/10.3390/land3041251
Lassus, B. (1991). Les continuitès du paysage. *Urbanisme, 250*, 64–68.
Lassus, B. (1998). *The landscape approach*. Penn University of Pennsylvania press.
Leone, A. (2009). *Riflessioni sul paesaggio*. Aracne.
Leong, S. T. (1998). Readings of the attenuated landscape. In *Slow space* (pp. 186–213). Monacelli.
Lerup, L. (1994). Stim and dross: Rethinking the metropolis. *Assemblage, 25*, 82–101.
Lettini, L., & Maffei, D. (1999). *La rappresentazione del Paesaggio*. Tirrenia Stampatori.
Lowenthal, D. (2015). *The past is a foreign country—Revisited*. Cambridge University Press.
Luginbühl, Y. (2009). Rappresentazioni sociali del paesaggio ed. evoluzione della domanda sociale. In *Di chi è il Paesaggio?* CLEUP.
Luginbühl, Y. (2017). Paysage et démocratie. In *Dimensions du paysage: Réflexions et propositions pour la mise en œuvre de la Convention européenne du paysage* (pp. 243–283). Conseil de l'Europe. https://doi.org/10.3917/europ.coll.2017.02.0243
Lynch, K. (1990). *Wasting away*. Sierra Club Books.
Margiotta, S., & Sansò, P. (2017). Abandoned quarries and geotourism. An opportunity for the salento quarry district (Apulia, Southern Italy). *Geoheritage, 9*, 463–477. https://doi.org/10.1007/s12371-016-0201-4
Marini, S. (2008a). *Architettura parassita. Strategie di riciclaggio per la città*. Quodlibet.
Marini, S. (2008b). Spazi Bianchi. Progettare lo scarto. In *L'architettura e le sue declinazioni*. Ipertesto.
Marini, S. (2010). *Nuove terre. Architetture e paesaggi dello scarto*. Quodlibet.
Marini, S., & Santangelo, V. (2013). *Re-Cycle Italy. Nuovi cicli di vita per architetture e infrastrutture della città e del paesaggio*. Aracne.
Martinet, J. (1984). Le paysage: Signifiant et signifié. In *Lire le paysage, lire les paysages, act du colloque des 24 et 25 novembre 1983*. CIEREC.
Matteini, T. (2009). *Paesaggi del tempo: Documenti archeologici e rovine artificiali nel disegno di giardini e paesaggi*. Alinea.
McDonough, W., & Braungart, M. (2002). *Cradle to cradle: Remaking the way we make things*. North Point Press.
McHarg, I. L. (1969). *Design with nature*. American Museum of Natural History.
Meyer, E. K. (2017). Uncertains parks: Disturbed sites, citizen and risk society. In *Large parks* (pp. 58–85). Princeton Architectural Press.
Miccoli, S., Finucci, F., & Murro, R. (2014). Social evaluation approaches in landscape projects. *Sustainability, MDPI, 6*, 7906–7920. https://doi.org/10.3390/su6117906
Mitchell, J. (2010). Rapid Re(f)use: 3-D Fabricated Positive Waste Ecologies. *Wiley Online Library*. https://doi.org/10.1002/ad.1172
Nabarro, R., & Richards, D. (1980). Wasteland: A Thames Television report. Thames.
Naveh, Z., & Lieberman, A. S. (1994). *Landscape ecology theory and application* (2nd ed.). Springer Science + Bussiness Media.
Nielson, T. (2002). The return of the excessive: Superfluous landscapes. *Space and Culture, 5(1)*, 53–62.
Northam, R. M. (1971). Vacant urban land in the American city. *Land Economics, 47*(4), 345–355.

Olin, L. (1988). Form, meaning, and expression in landscape architecture. *Landscape Journal, 2*, 149–168. https://doi.org/10.3368/lj.7.2.149
Paulhan, F. (2016). *L'Esthetique Du Paysage*. Librairie Felix Alcan.
Pierre, G. (1982). René Louis de Girardin, De la Composition des paysages, suivi de Promenade ou itinéraire des jardins d'Ermenonville. *Annales de Géographie, 505*, 387.
Pineiros, E. (2016). *Terrain Vague and the Urban Imagery*. [Doctoral Thesis]. University of Nottingham.
Pinto-Correia, T. (2000). Landscape identity, a key for integration. In *Landscape- Our home* (pp. 145–150). Indigo-Zeist.
Pinto-Correia, T., & Ramos, I. (2000). As identidades locais em espaço rural: As tradicionais e as novas funções da paisagem rural. In *Identidades Locais e Globalização*.
Poli, D. (2002). *Progettare il Paesaggio nella crisi della modernità. Casi, riflessioni, studi sul senso del paesaggio contemporaneo*. All'insegna del Giglio.
Raffestin, C. (1978). Du paysage à l'espace ou les signes de la Géographie. *Hèrodote, 9*, 90–104.
Raffestin, C. (2012). Space, territory, and territoriality. *Environment and Planning: Society and Space, 30*, 121–141.
Rathje, W., & Murphy, C. (2001). *Rubbish: The archeology of garbage*. University of Arizona Press.
Ribeiro, O. (1998). *Portugal, o Mediterrâneo e o Atlântico*. Livraria Sá da Costa Editora.
Ribeiro, O., & Lautensach, H. (2009). *Geografia de Portugal* (Vol. 1). Edições João Sá da Costa.
Salvini, R. (1964). *All the Paintings of Giotto*. Hawthorn Books.
Secchi, B. (1989). *Il vuoto. Un progetto per L'urbanistica*. Einaudi.
Secchi, B. (1993). Un'urbanistico di spazi aperti/For a town-planning of open spaces. *Casabella, 597–598*, 116–117.
Sénécal, G., & Saint-Laurent, D. (1999). Espaces libres et enjeux écologiques: Deux récits du développement urbain à Montréal. *Recherches Sociographiques, 40*(1), 33–54. https://doi.org/10.7202/057243ar
Sénécal, G., & Saint-Laurent, D. (2000). *Les espaces dégradés: Contraintes et conquêtes*. Presses de l'Université du Québec.
Shama, S. (1995). *Paesaggio e Memoria*. Mondadori.
Simmel, G. (2007). The philosophy of landscape. *Sage Journals, 14*(7–8). https://doi.org/10.1177/0263276407084465
Silva, C. F., & Esteves, L. D. (2015). The Mounds of Estremoz Marble Waste: Between refuse and reuse. In *Waste 2015—Solutions, treatments and opportunities* (pp. 305–310). CRC Press.
Simão, J., Freire, A. C., & Farto, H. (2016). *Caracterização de rejeitados de escombreiras de pedreiras de rocha ornamental para aplicação em camadas não ligadas de pavimentos rodoviários* (pp. 1–9).
Smithson, R. (1979). Incidents of mirror-travel in the Yucatan. In *The writings of Robert Smithson* (pp. 94–95). New York University Press.
Smithson, R. (1996a). A sedimentation the mind: Earth projects. In *Robert Smithson: The collected writings* (pp. 110–111). University of California Press.
Smithson, R. (1996b). Sites and settings. In *Robert Smithson: The collected writings*. University of California Press.
Soares, B. (1982). *Livro do Desassossego*. Ática.
Spinarelli, G. (2016). *La riscrittura dello scarto. Il caso dell'ex cementificio Italcementi di Albino*. Milan Politecnic.
Stalker. (1996). *Manifesto*. www.osservatononomade.net/tarkowsky/mamfesto/mamfestmg.htm
Talento, K., Miguel, A., & Kullberg, J. C. (2018). The metamorphosis of the landscape: The adaptive reuse of marble waste. In *The fifth international conference on architecture and built environment with AWARDs S.ARCH 2018 conference proceeding*. (Get It Published).
Tan, F., Jiao, Y.-Y., Wang, H., Liu, Y., Tian, H., & Cheng, Y. (2019). Reclamation and reuse of abandoned quarry: A case study of Ice World & Water Park in Changsha. *Tunnelling and Underground Space Technology, 85*, 259–267. https://doi.org/10.1016/j.tust.2018.12.009

Tarpino, A. (2012). *Spaesati. Luoghi dell'Italia in abbandono tra memoria e futuro*. Passaggi Einaudi.
Tilley, C. Y. (1994). Space, place, landscape and perception: Phenomenological perspectives. *A phenomenology of landscape: Place, paths and monuments*, 27–30.
Tsolaki-Fiaka, S., Bathrellos, G. D., & Skilodimou, H. D. (2018). Multi-criteria decision analysis for an abandoned quarry in the Evros Region (NE Greece). *Land, MDPI, 7, 43*. https://doi.org/10.3390/land7020043
Tufnell, B. (2006). *Land art*. Tate.
Turri, E. (2003). *Il paesaggio degli uomini. La natura, la cultura, la storia*. Zanichelli.
Twigger-Ross, C., & Uzzell, D. L. (1996). Place and identity processes. *Journal of Environmental Psychology, 16*, 205–220.
Ulubeyli, G. C., & Artir, R. (2015). Properties of hardened concrete produced by waste marble. *Procedia Social and Behavioral Sciences, 195*, 2181–2190.
Usher, M. B. (1999). *Landscape character. Perspectives on management and change*. The Stationery Office.
Venturi Ferraiolo, M. (2006). *Paesaggi rivelati. Passeggiare con Bernard Lassus*. Guerini e associati.
Veríssimo Serrão, A. (2011). *Filosofia da Paisagem, Uma Antologia* (2nd ed.). Centro de Filosofia Universidade de Lisboa.
Viale, G. (2000). *Un mondo usa e getta. La civiltà dei rifiuti e i rifiuti della civiltà*. Feltrinelli.
Vosloo, P. (2018). Post-industrial urban quarries as places of recreation and the new wilderness—A South African perspective. *Town and Regional Planning, 72*. https://doi.org/10.18820/2415-0495/trp72i1.4
Waldheim, C. (2016). *Landscape as urbanism: A general theory*. Princenton University Press.
Washer, D. (2003). Learning from European trans-frontier landscapes. *Second Meeting of the Workshop for the Implementation of the European Landscape Convention, T-FLOR, 3(2003), 12*.
Waterman, T. (2009). *The Fundamentals of Landscape Architecture*. AVA Publishing.
WDe Solà Morales, I. (1995). Terrain Vague. In Anytime (pp. 118–123). MIT Press.
Weilacher, U. (1999). *Between landscape architecture and land art*. Birkhäuser.
Weilacher, U. (2005). *In gardens. Profiles of contemporary European landscape architecture*. Birkhauser.
Zagari, F. (2006). *Questo è paesaggio 48 definizioni*. Mancuso Editore.

Chapter 3
The Rehabilitation of Quarries

How to make old things new, how to see something common and banal in a new and fresh way is the central problem

Laurie Olin

3.1 Introduction

Analysis of the concept of quarries' rehabilitation, spanning from a utilitarian to an ecological perspective, and elaboration of a general schematization illustrating the different reuse typologies of a quarry.

The Earth's crust is subjected to constant mutations through physical and chemical fragmentations. Between 2000 and 2005, two scientists, E. Stoermer and P.J. Crutzen, developed a theory related to a new, ongoing geological era, defining it as Anthropocene. In this epoch, which started in 1784 with the invention of the steam engine, humanity is considered to be a real geological agent, due to the strong impact on the atmosphere and on Earth's surface (Fuentes & Baynes-Rock, 2017). This acceleration of the human impact was a result of the introduction of new technologies (Eldredge & Pievani, 2010). Humans were the protagonist of the Anthropocene epoch, in which they were also able to understand their mistakes and to intervene, through political strategies, for control of the territory (Crutzen, 2005). The extraction industry falls entirely within anthropic activities, which can completely alter the environment. This industry accompanied humans from the prehistoric era when they dug rock to fabricate tools and obtain construction materials and pigments. Therefore, they usually acted on Earth, modifying and adapting it to their necessities, regardless of the consequences in the aesthetic, ecological, and environmental sectors (Aliberti, 2015; Goudie, 2018; Marcantonio & Fuentes, 2020; Tarolli et al., 2019).

This chapter provides an exhaustive report regarding the concept of quarry rehabilitation, after degradation from industrial activity. On the one hand, practical interventions have increased and multiplied worldwide; on the other hand, there is

K. Talento, *Sustainable Buildings from Marble Waste Heaps*,
https://doi.org/10.1007/978-3-031-72146-5_3

an architectural gap in the theoretical and scientific sectors. The concept of quarry rehabilitation has been analyzed to describe specific examples of regenerations in particular areas, to evaluate the environmental impacts generated by quarries, explain the main techniques of reconversion, and illustrate the recovery methodologies associated with dust and wastes in the construction industry (Chetia & Sridharan, 2016; Galetakis & Soultana, 2016; Kirtan & Siddharth, 2015; Langer, 2001; Legwaila et al., 2015; Montañez-Escalante et al., 2005). In these examples, the possibility of having a general view of the topic, in the architectural sector, was lacking. For this purpose, the research proposes the current work as a useful instrument to improve the scientific knowledge surrounding the topic of quarry recovery. In line with the innovative principles of sustainability, this research explores the transformation possibilities aimed at reintegrating these areas in ecological and territorial metabolisms, re-establishing the material flows that occur between humanity and nature. As will be shown in the next sections, this objective has been achieved selecting, among countless examples of quarry reconversions, the most crucial ones, deepening and classifying them. In this regard, the research develops a useful schematization, in which the simple methods of the conversion have been classified to give a general and critical view of the existing rehabilitation techniques.

"Starting from this schematization, will it be possible to investigate and derive new methods of rehabilitation?"

This is the key question discussed in the present chapter to understand how abandoned and industrial territories can be transformed into unique places while minding the sustainability concept. Although this study is firmly attached to the environmental design field, it benefits from an interdisciplinary nature that enriches the work. Section 3.2. discusses a brief history of the notion of quarry reuse. Section 3.3. describes the new value of the industrial landscape. These two sections provide the essential theoretical basis for the following ones. Section 3.4. addresses some criteria for the reconversion of former quarries, by means of the referred schematization. Moreover, the research methodology, the scientific schematization, and the discussion of quarry reconversion examples are provided at the end of the section. Section 3.5. explains the concept of reuse in the contemporary era highlighting the current quarry trends. Finally, Section 3.6. draws the main conclusions.

3.2 A Brief History of Quarry Reuse: From a Utilitarian to an Ecological Approach

Quarries represent remarkable resources from the ground. There are several historical examples where there has been a reuse of these areas for utilitarian purposes. Recently, due to the growing sensibility toward environmental issues, the possibility of ecological reuse flanks the utilitarian purposes. To briefly summarize the historical concept of the reuse of the extraction places, the range of funeral sites is to be considered; the catacombs represent one of the most significant cases in this sector.

The term "catacomb" was born in Rome, and it derives from "*catacumbas*," namely "at the cavity" *for* the S. Sebastiano funeral complex in the Via Appia located inside an ancient stone quarry. The catacombs occupied a considerable number of artificial and abandoned cavities (such as tuff, clay, and stone quarries), and, in the Roman epoch, they became interesting places for the burial of the dead, to celebrate religious rituals, and to escape from persecution (Aliberti, 2015). For this type of reuse, the quarries were subjected to minimal interventions of safety and, only in certain cases, they were remodeled and frescoed. The Priscilla catacombs, in Rome, are a beautiful example based on this concept (Fig. 3.1) (Cifani, 2008).

It is significant to note how many cities took possession of quarries to obtain construction materials. The city of Naples, for instance, has exploited, since the Roman era, the subterranean tuff quarries, to convey water from the Serino River. This operation permitted the creation of tanks, linked to each other, for the water supply of the city until the seventeenth century, before being emptied (due to the construction of a new aqueduct) and reused as cemeteries and shelters (Fig. 3.2). In Paris, *les Carrieres,* Roman limestone quarries, were turned into town ossuaries to resolve the problem of inappropriate sepulchres in the neighboring cemeteries.

In Syracuse, Italy, there are several examples of Latomie, ancient stone quarries, where the Greeks extracted all the necessary material for the realization of temples, roads, walls, and defensive works. Some of these pits were also used for the imprisonment of slaves and criminals and, in more recent times, they became dwellings for the humblest classes until now, where they have been transformed into gardens (Aliberti, 2015).

Fig. 3.1 Priscilla Catacombs, Rome. (Wikipedia, 2020)

Fig. 3.2 Le Fontanelle Cemetery, Naples. (Wikipedia, 2020)

With tragic ends, one remembers the Mauthausen concentration camp in Austria (Fig. 3.3). During the First World War, the Austrians opened the first camp for war prisoners at the quarry of Wiener-Graben, previously used from the exploitation of Viennese granite used for the pavement of Vienna roads. During the Second World War, this camp became one of the cruelest death-camps during the Nazi regime. In these years, countless discarded quarries were rehabilitated and reused as air-raid shelters, as happened in Colleferro, Italy (Fig. 3.4). Many people who settled in these refuges also made the space more comfortable, blanketing the walls with lime for hygienic and lightning issues. After Liberation, the air-raid shelters were partially reutilized as mushroom gardens. The same spaces were also present in Naples, with the difference of being more equipped with the electricity network, toilets, furniture, and distinct areas for the first and second classes. These real examples are essential to highlight the exclusive utilitarian approach for the reuse of the extractive sites, without any environmental and ecological preoccupation or any landscape meaning.

The landscape approach was introduced during the Renaissance age in Florence, with the recovery of quarries to be allocated to gardens. The Boboli Garden is today the biggest green area inside the walls of Florence, based on an ancient spot already anthropized by agricultural and extractive activities (Fig. 3.5). The Boboli hill was used as "*pietraforte*" quarry to provide material for the first paving of the city streets, and the construction of the town palaces, such as the Pitti Palace. The same Florence model was adopted in Paris with the project for the Buttes-Chaumont Garden, in Vancouver with the realization of the Queen Elizabeth Park, and in Ischia

Fig. 3.3 Mauthausen concentration camp, Austria. (Wikipedia, 2020)

(Italy) with the creation of La Mortella gardens (Fig. 3.6). Each case was based on the transformation of ancient quarries into beautiful parks with aesthetic and environmental qualities (Crutzen, 2005). According to the explanations previously mentioned and to the reality of our mutated surroundings, the presence of an industrial landscape, and, specifically, of a quarry on the territory, has always led to alterations of the land and to the urgency and the need to reactivate it.

Fig. 3.4 Air-raid shelters, Colleferro, Italy. (Wikipedia, 2020)

3.3 The New Value of the Industrial Landscape

The industrial landscape represents an atypical landscape category, as it extends the landscape perception to places, such as those of production, which are born and developed without any aesthetic purpose from the people who made them. Nevertheless, the industrial spots, once finished in their production activity, reveal unsuspected aesthetic qualities, useful for possible operations of conservation and valorization of these sites. The importance of the industrial landscape is related not only to the new aesthetic associated with places of work but also to the conservation of the cultural value of the place for the allocation of a different and original function, compatible with the society of our time.

The extractive areas, in a state of abandonment, compose a particular landscape in which the technological heritage is intertwined with the environmental one, creating a patrimonial pluralism that requires innovative forms of protection and valorization. This conception was reached only at the end of the 1980s. Previously, the "voids" left by de-industrialization were synonymous with tabula rasa, which erased all the productive traces jointly with the memory of a past by forgetting. On the contrary, between the 1980s and 1990s, the conservation and valorization of industrial sites were considered as the growth of the cultural heritage of the cities to which they belong (Preite, 2017). On par with the ancient city squares, in the

Fig. 3.5 Boboli Gardens, Florence. (Wikipedia, 2020)

industrial areas, it is also possible to trace a consistent depth of historical stratification. They are deposits of a more recent history, no less intense: histories of fights, hard work conditions, solidarity between workers, and antagonism for the industrial discipline. The depth of the memory of the workplaces is undeniable, and it represents one of the main components of the immaterial heritage.

The industrial landscape is constituted as an "anti-landscape" rather than traditional; the traditional landscape is a landscape of continuity based on traces of the long-standing history of the territory. The industrial landscape is also the landscape of discontinuity, as its appearance indicates a breaking point, compared to the previous history, and a new beginning in the events of the territory. The industrial landscape is the landscape of the metamorphosis of the values that could be transformed into a place rich in memory with aesthetic dimensions (Preite, 2017). This transformation implies the need for urban regeneration to rethink the economy, the function, and the destination of use of the abandoned industrial sites.

The recovery of abandoned quarries does not resolve through a bland "masking" of the excavation walls but through the reappropriation of the cultural aspects of the territory. In this way, the definition of a different cultural horizon is necessary to reinsert the dismissed quarries into a new circuit, formed by common uses, where the possibility to reinvent the anthropic landscape and its relationship with the natural one exists. Designing the profound spaces of the landscape refers to the modern notion of hypogeum, able to produce vastly different architectural articulations and an innovative association with the territory and its resources (D'Amato, 2005).

Fig. 3.6 La Mortella Gardens, Ischia, Italy. (Wikipedia, 2020)

Even if the activity of digging for rocks is as ancient as the presence of humanity on Earth, the question of territorial regeneration, focused on the recovery of the dismissed productive sites, is one of the central arguments of the contemporary debate. The considerable amount and extent of the former extraction systems, located under our urbanized soils, make their protection, recovery, and reuse urgent. Their consolidation and securing are necessary to prevent the eventual phenomenon of instability (Tsolaki-Fiaka et al., 2018). Most often, these cavities reveal notable qualities and spatial values, and they represent a typical porous sub-stratum, which crosses the anthropized territory, claiming the need to have a new use.

There are different ways to confer a new meaning to these unused and porous sites. For example, reorganizing the hypogeum cavities into a museum, for visits and thematic exhibitions; through the necessity to allocate useful services for the city; or for structural and consolidation needs. Regardless of the various orientations, the recovery intervention will have a utilitarian valence, being productive, resolute, and a symbol of transformation, innovation, and progress. Recovering and reconverting a former extractive hypogeal site means to recognize the opportunities and, at the same time, the limits offered by the quarries, the nature, and the characteristics of the excavated material (Shi et al., 2020). The regenerative force exercisable by these marginal voids is becoming the strong point of the so-called new ecologies, which uses the existent heritage at the center of the project for the contemporary city. The intention is that of generating new life cycles from the devastated territory, which has suffered the most aggressive forms of anthropic degradation (Dell'Aira et al., 2015). This intervention produces new ecosystems characterized by a substantial complexity that get to live only if they are maintained, incremented, and assisted in their constant mutations (Magnaghi, 2010).

Therefore, with rehabilitation projects, the community can accept the dismissed areas for a multitude of collective uses, taking the opportunity to reorganize and decentralize the urban poles for a rebalancing of the anthropic cargo on the territory (D'Amato, 2005). Through the reconversion of former extractive sites, the territory also acquires other opportunities in both the economic and ecological sectors. In this regard, the quarries are transformed from "Gray Infrastructures" to "Green Infrastructures." The first category is related to the works drafted for monofunctional specific uses (such as bridges, highways, and dams), mainly constituted by manufactured materials, contrasting with natural systems. On the contrary, the second category includes natural or seminatural multifunctional areas, with a rural or urban character, which are designed to increase the liveability and the environmental values of the local communities. The Green Infrastructures, by encouraging the use of natural capital, are conceived to ecologically equilibrate areas and territories, by providing a wide range of ecosystem services (Fig. 3.7) ([MEA] Millennium Ecosystem Assessment, 2005).

3.4 Some Useful Criteria for the Reconversion of Former Quarries into Renewed Places

3.4.1 The Morphology of the Territory and Destinations of Use

Most often, the obligation to confer a renovated use and meaning to the degraded extractive sites implies transformation operations for the place. These metamorphoses can require diverse types of interventions, which span from minimal measures to considerable changes (such as filling, reforestation, and new constructions). Typically, different parameters confront each other in these complex procedures, such as the quarry typology, the dimensional and formal character, the security of the areas, and the ecology (Tang et al., 2011). Any prediction of the transformation

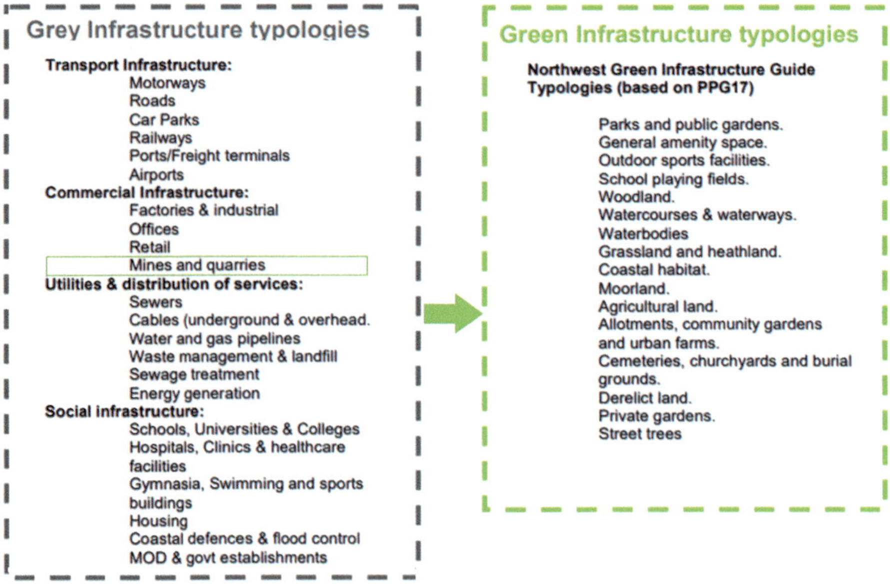

Fig. 3.7 Component parts of Gray and Green infrastructures. Source: NENW, 2009 (Fabbri, 2007)

presupposes, on the one hand, a comparison with analog cases to identify similarities and to define concepts and, on the other hand, a close reading and analysis of the site and its context to evaluate and design solutions with a strategic approach (Aliberti, 2015).

Regarding the transformation processes of a quarry, there are some essential elements to keep in mind. One is the morphology of the territory, which depends on the typology of the extraction site (Dávid & Patrick, 1998). If one considers the quarries as patches within a landscape mosaic, one notes that they have identifiable geometric features. Since humanity produced them, these characteristics are different from those of natural origin (Fig. 3.8). Entropy leads to casual shapes, convoluted and with irregular perimeters; instead, the forms made by humans are normally evident (Fabbri, 2007).

In this respect, the question related to the borders of a quarry is crucial to understanding the difference between the artificial and the natural patches. In the first case, most of the shapes made by human activity present straight lines, due to the need for managing, regularizing, and controlling that part of the territory (tracks, terraces, and large steps). Between the other variables of the form parameters, the aspects linked with the compactness of the borders also gain significant consideration (Fabbri, 2007). The compact configurations are determinant to maintain the resources of a quarry, decreasing at the same time, the movement outward because of the limited quantity exposed through the margins. On the contrary, the articulated forms are crucial in increasing this interaction with the surroundings.

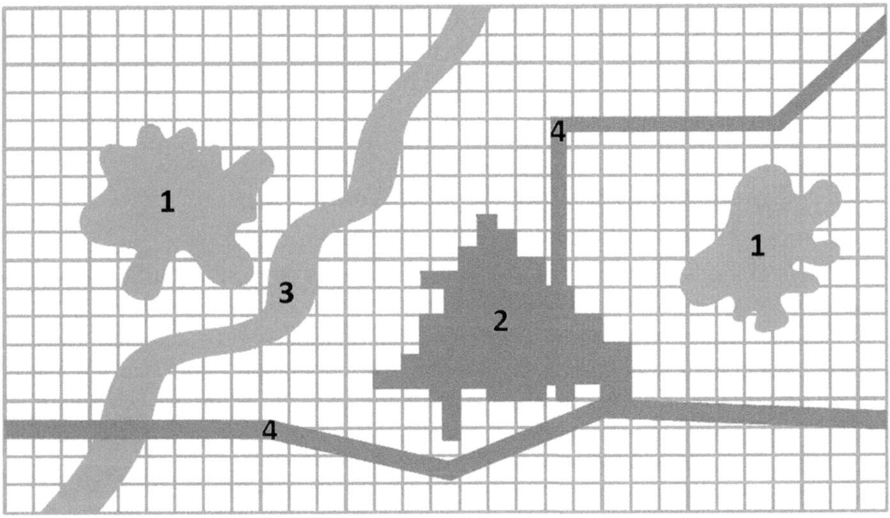

Fig. 3.8 Environmental mosaic scheme composed of different patches and corridors. *1*. Natural patch, *2*. anthropic patch, *3*. natural corridor, *4*. anthropic corridor. (Source:Fabbri, 2007)

After the classification of the primary criteria of form for the quarry rehabilitations, the research investigated the design and the transformative possibilities of the extractive areas regarding the "destination of use" parameter. The recovery typologies of the abandoned quarries can be schematized according to the following purposes: naturalistic, recreational, cultural, educational, and productive.

The naturalistic recovery of a quarry is an intervention method based on vegetational and forest recovery, through the realization of green areas. It provides the creation or the reconstitution of the natural features of the quarry zone, in connection with the surrounding environment (Rosiere et al., 1989). This type of recovery aims to limit the morphological effect caused by the excavation (Mouflis et al., 2008), and it proposes to recreate a habitat, as similar as possible, to the natural one. This category of recovery requires an in-depth ecological analysis of the territory to intervene with adequate technical and vegetational choices. This category also includes all those lands, which, after their exploitation, acquired particular geological conditions that need protection. Examples include the quarry or mining areas (e.g., in Sardinia and Apulia), which have assumed the connotation of "geo-mining parks," a symbol of an important cultural heritage (D'Amato, 2005).

The recreational rehabilitation, jointly with cultural and educational goals, aims to realize distinct spaces, intended to receive services and equipment with recreational and educational functions: touristic equipment, museums, galleries, musical and sports activities (Damigos & Kaliampakos, 2003), and leisure facilities.

The productive replacement includes agricultural and touristic recovery. In the first case, the scope is to take back the dismissed quarry to its original condition of an agroecosystem. The second case consists of making the territory more attractive to encourage visits from tourists, giving a renewed meaning to the place. The preferable destination for the productive options includes areas rich in population density and residential and productive settlements, or areas that are border places with particular aspects (Talento et al., 2019a).

3.4.2 Typologies of Reuse: Projects Comparing

The considerations aforesaid are an important premise for the identification of the strategies for the requalification of the extractive areas. Deepening this study, a synopsis of the distinct typologies of quarry rehabilitation is provided. The research elaborated a general schematization in which the fundamental approaches of recovery have been classified, not to simplify the concept but to offer an overview of the existing possibilities and rehabilitation techniques. The typologies of recovery are subdivided into two macro groups:

Remodeling: It includes operations of filling, modification, and insertion.
Not remodeling: It does not entail substantial alterations of the excavation.

Concerning the first macro group and the activity of "**filling**," there is a distinction based on the diverse types of materials used, which can result in specific and peculiar conformations. For instance, if the filling matter is the water, the possible configurations are those typical of lakes, reservoirs, and wetlands, where the flat surfaces will prevail, rather than the protruding volumes. One of the most famous examples is the **F60, a vast mining area in Lusatia** (Brandenburg) (Kus et al., 2019), formed by extensive open-air deposits of lignite (brown coal). These deposits give the region the nomenclature of "Energy district" (Morton & Müller, 2016), in which power stations, briquettes factories, coke ovens, and other establishments are located. After political changes, most of the quarries were closed and the government allocated special funds for the rehabilitation of these territories, through the institution of a state company, the *Lausitzer und Mitteldeutsche Bergbau-Verwaltungsgesellschaft* (LMBV). This association preserved the industrial heritage and created a new way to reuse the brownfield (P. Langer, 2019; Talento et al., 2019b). The main characteristic of these post-extractive sites is the presence of new water surfaces, similar to lakes, due to the filling of the ancient pits (Blanchette & Lund, 2016) for recreational and sports activities (De Poli & Incerti, 2014; Deshaies, 2018) (Fig. 3.9a).

In other situations, solid materials, such as the ground, represent the infill that expands the possibilities of reshaping and recreating morphological continuity from the margin zones. Belonging to this classification, the project of the **Royal Botanic Gardens Victoria** in the city of Cranbourne (Australia) was born of a former sand quarry. The goal of the architectural project was to create a large botanic garden, with recreational and educational purposes, to display the different Australian micro-landscapes (Bull, 2006). The garden was designed by Taylor Cullity (Lethlean firm) in collaboration with the botanist Paul Thompson, who is a massive collector of Australian flora. It was developed along a water path, which metaphorically

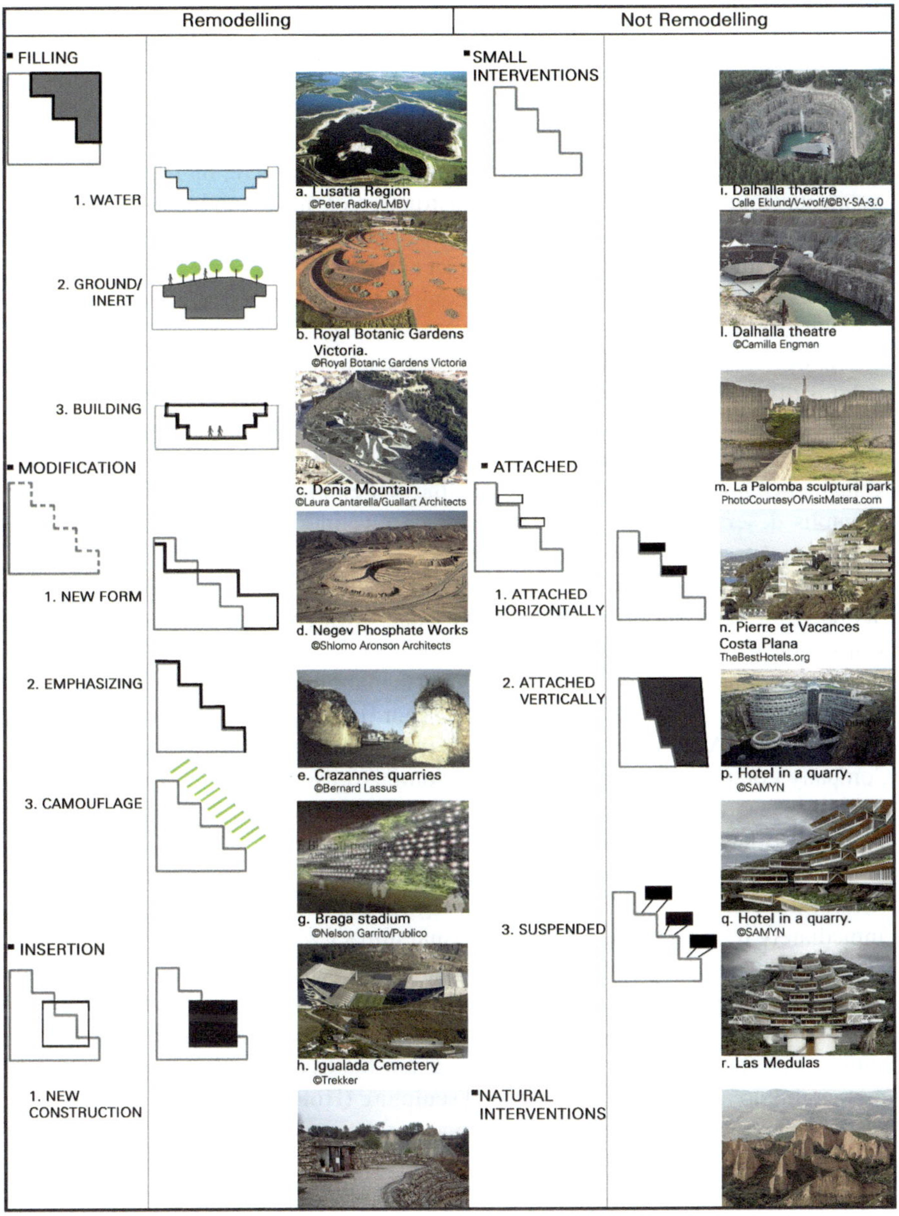

Fig. 3.9 Schematization of existing rehabilitation typologies for quarries [Katia Talento, 2020. Partially elaborated from (Aliberti, 2015)]

crosses the arid landscape of the Australian hinterland. One passes from the dry riverbed to artificial lakes up to the coast allowing the visitor to ideally retrace the water travel of the continent. The recovery project has reshaped the ground flows

(Fig. 3.9b) and the realization of wet zones, alternate with gardens composed of 17,000 species of plants, has turned this place into a premier site dedicated to Australian flora and biodiversity (Aliberti, 2015).

In addition to the typologies previously mentioned, the research adds another one that includes the method of the "constructed void" for the filling technique. This is the case of the un-realized project **Denia Mountain** where, on the one hand, the architects simulated the volumetric reconstruction of the hill and, on the other hand, they "involved" the empty space to allocate for specific functions, such as parking, hotel, and auditorium. However, the project aimed to highlight the artificiality, using a different treatment of the surface, composed of terraced gardens and ramps (Guallart Architects, n.d.) (Fig. 3.9c).

The operation of "**modification**" falls in the second subcategory of remodeling, and it is divided into three distinct typologies: new form, emphasising, and camouflage. "**New form**" refers to new aesthetic, visual, and ecological configurations, designed to reduce the phenomena of degradation through land movement or rock reshaping (Aliberti, 2015). Examples are the **Negev Phosphate Works**, in the homonymous desert in Israel, where a major mining industry for the exploitation of phosphates, used for fertilizer, is located.

For decades, work has removed these materials by open-pit mining. The result has been the creation of a desolate landscape of huge holes with adjacent waste heaps some 40 m high (Shlomo Aronson Architects, 2020). However, the vast scale of the tipped material was utterly foreign to the landscape of the Zin Valley. This is an impressive desert wadi landscape of crescent Hogback hills, which is also subject to flash floods. After the extraction of the material, in 1990, the Negev Phosphates Company commissioned the architect Shlomo Aronson to design a more environmentally sensitive approach to opencast mining. He devised methods of filling in the excavations, left by mining with the spoil material.

The strategy adopted to resolve the problems has three main elements: First, the conservation of the wadi involves breaking the mounds of waste in two and, this immediately reduces the scale of their visual impact while also conserving the side slopes of the Zin Valley. Second, new landforms reflect the existing topography. Finally, the impact and the scale of the earthwork were reduced by creating 10-meter-high setbacks instead of 30–40 m. In this industrial land, the architect acted as a sculptor and the result was the creation of a new landscape, with a new morphology, comparable to a gigantic environmental sculpture (Holden, 2003) (Fig. 3.9d).

The typology of "**emphasizing**" concerns actions of sculptural remodeling; employing cuts and bevels, these landforms exalt the scenographic features of the form (Fig. 3.9e). An interesting example is the landscape intervention of Bernard Lassus in the **Crazannes quarries**, along with traits of the French highway. In this rest area (de Wit, 2016), some stone masses have been removed or reshaped to open the original views and scenarios in depth, using contrasts between the masses and voids, the lights and shadows, the green of the vegetation, and the white of the stone. There is a work of reinterpretation of the places and functions that bring out several rocks of the forgetting quarries, covered by vegetation. In the project, the maintenance of the fern wood represents the vegetal drawing more evidently. The sequences of excavations, invaded by plants, form a vast romantic landscape, in which the succession of distinct points of view unifies the former quarry to the road infrastructure. In this way, the driver, along the highway, recognizes the meaning, the form, and the memory of the place (Conan, 2012).

The "**camouflage** "method, in conclusion, aims to hide undesired surfaces or substances through technological solutions (Bellini, 2013). **Biovail** is a significant project in which minimal technological supports (such as canvas ropes, nets of natural fibers, and recycled tubes) were overlapped to the rock to treat the wounds inflicted on nature (Cesarin & Fornari, 2010) (Fig. 3.9f).

Finally, the **insertion** operation is a part of the last typology of the remodeling phase. It comprises interventions of new constructions partially inserted in the volume of the quarry, causing excavations of subtraction and adaptation. Among the most suggestive examples of this category is the **Braga Stadium** in Portugal, made by the architect Souto de Moura, and the **Igualada Cemetery** in Spain, of Enric Miralles and Carme Pinos. The Braga stadium (Fig. 3.9g) is the result of two major decisions. The first emerged at the particular moment when the architect visited Monte Castro, in Portugal, recognizing the old and deactivated quarry as a perfect place for the stadium implantation (Architeam., 2013; Archzine, 2014; Fidalgo, 2017). He chooses a landscape already transformed by men, instead of an untouched one (Salema, 2005). The second decision was to perceive the stadium as a large infrastructure that made a functioning machine. The construction is within the sports park of Dume, on the northern slope of the Castro Mountain, and its strategic position avoided the construction of a dam in the reservoir of the valley (Maza, 2004).

Due to the great importance of football in the sporting world and its power of entrainment, the architect chose to realize only two wide and contrasting seating steps. The first was embedded in the rock face; the other one emerged from the ground reconstructing a basin, closed on one side by a mountain crest and opened, on the other side, to the valley's scenic view. Concerning the coverage of the stadium, the original idea showed a long roof similar to that of the Siza Expo pavilion (De Poli & Incerti, 2014), but, subsequently, the firm adopted a different solution taking inspiration from the Inca bridges in Peru. For this reason, with a height of 40 m, the stadium is formed by two square suspended top floors having the same inclination (Diord et al., 2017).

The **Igualada Cemetery** was designed to be a place of reflection and memories (Fig. 3.9h). The authors understood it as the "city of the dead" where the dead and the living coexist together in spirit. Therefore, the project is also a place to rest and reflect on the solitude and serenity of the Catalonian landscape of Barcelona. The architects conceptualized the poetic idea of a cemetery for the visitors to comprehend and accept the cycle of life as a link between the past, present, and future. Incorporated in the Catalonian hills, the cemetery is an earthwork located inside an arid river valley (Cook, 2016) that merges with the landscape as if it were a natural feature of it. Thus, the project was conceived as an element that transforms the surrounding landscape and as a metaphor for the river of life. A processional way descends from the entrance, where crossed, rusting, steel poles double as gates, and this is comparable to the crosses at Calvary. The pathway is marked by replicable concrete loculi, which form supporting walls and by the concrete floor. The intention was to bring the bereaved down into the landscape to a "city of the dead" to make it communicate with the living (Souza, 2013).

Concerning the recovery category of the **insertion**, and regarding the two projects the research just described, the architects applied morphological grafts with the partial inclusion of new volumes within the edge of the quarries. The second macro category describes the solution of "**not remodeling**" for all the places, which, although deformed, present special features of form. For all these cases, only small interventions are performed, related to the securing of the sites and the excavation. In this case, the rehabilitation of a quarry is realized without any morphological modification, taking advantage of the form peculiarities of the site.

Several examples refer to places destined for sports-related, artistic, and cultural activities. One cites the case of the sculptural **park La Palomba**, in Matera, or the case of the former quarry of Rättvik, in Sweden, transformed in the **Dalhalla theater**, in which the architects took advantage of the natural configuration of the basin, using its considerable acoustic capabilities for the promotion of recreational activities, concerts, and events during the summer season. Dalhalla is an amphitheater, with descending benches down to an emerald lake in the flooded part of the pit (Fig. 3.9i,l) (Rybnikovaa et al., 2017).

The sculptural park La Palomba, located in the historical natural archaeological park of the Matera rock churches in Italy (Fig. 3.9m), symbolizes an interesting anthropological work (Fallacara et al., 2016). The artist Antonio Paradiso decided to rehabilitate the former industrial space, transforming it into an artistic and cultural area, enriched by collective expositions and temporary exhibitions. The sculptor creates its works according to the practice of the artistic residences with a stay in the "stone city," in direct contact with the territory. The new site develops in the six hectares of an ancient tuff quarry material that has for centuries characterized the architectural history of Matera city (Aliberti, 2015).

Within the category of "not remodeling," the "**attached**" is also considered, and it is declined in three variants: "**horizontally attached**," "**vertically attached**," and "**suspended**." These are measures, which, although do not include the morphological transformation of the excavated empty spaces through land movements, adopt the construction of buildings, paths, and facilities to resolve the metamorphosis of a quarry. Three representative projects embody these distinct approaches: regarding the first technique, the **Pierre et Vacances Costa Plana** in France of the architect Jean Nouvel represents a significant example (Fig. 3.9n). In this case, the architecture and the vegetation (composed of pine trees and agaves) cling to the steep surface of the quarry. It seems almost like a cliff, where the buildings appear with simplex and rhythmic volumes, "disappearing" into the rock and behind the flora.

Concerning the second approach, the **Intercontinental Hotel** in Shanghai of Jade architects, and associates, constitutes the best instance of a building constructed and attached to the vertical wall of the quarry. The studio described it as an "inverted ground-scraper," since it represents a tall construction (88 m deep) fixed at both ends having a 337-room hotel that descends 16 stories below ground level, within the former quarry. In this regard, the walls of the pit wrap the rooms, and a "glass waterfall" structure contains the building's observation lifts for access to all floors (Fig. 3.9o).

The **Hotel in a quarry** in China of the firm Samyn and partners symbolizes an edification closer to the third approach. The present project provided a 20-room hotel facility and a small open-air theater in an existing quarry. The 20-bedroom feature hotel at the upper level, in the form of individual cabins, is attached to the cliff and is accessible through a funicular lift and horizontal pathways (Fig. 3.9p,q). All these cases can be perceived as "foreign" objects; however, at the same time, they appear well integrated with the natural aspects and the surroundings.

Concurrent with all these artificial interventions, other typologies of recovery exist, dictated by the natural spontaneity of reconversion. This is a concept that is associated with the thought of the scientist James Lovelock, and it exalts the power of nature to destroy any form of degradation. According to the author, despite the excessively long periods, nature will restore the lost balance through spontaneous biological processes (Lovelock, 2006). This theory is demonstrated by several examples; one of them is the former gold mine, **Las Medulas**, in Spain (Fig. 3.9r). Other exceptional places are the canyons, ravines, or volcanic lakes that have become precious refuges of biodiversity, creating new ecosystems.

In this regard, natural recovery is a still used process in which re-vegetation relies on the presence of seeds and roots in the soils, transported from adjacent lands, through natural courses. This development happens through different stages, referred to as primary and secondary succession, which occur through the processes of species colonization, spreading, and replacement over time. In some cases, because of the limitations of soils in quarries, parts of the excavation could remain exposed, due to lack of vegetation cover. This aspect could provide important habitats for a variety of wildlife, as well as create opportunities for exhibitions of the industrial archaeology (Legwaila et al., 2015).

The notion of natural intervention concludes the schematization of the principal typologies of rehabilitation of quarries. From this systematization, the research describes the selected cases studies (a few of a wide range) to extrapolate the most relevant concepts of transformation, analyzing them from an architectural and landscape point of view. For the writing of this chapter, several references have been selected for the organization of the literature review and they have been searched for keywords, mostly related to the concepts of waste, waste reuse, industrial landscape, and quarry rehabilitation. The bibliography was then organized to continue to follow the intention of the theoretical part of the research, namely that of creating a fluent reading about the waste notion, from its origins to its contemporary mutations.

In recent years, numerous publications (Chetia & Sridharan, 2016; Galetakis & Soultana, 2016; Kirtan & Siddharth, 2015; W. H. Langer, 2001; Legwaila et al., 2015; Montañez-Escalante et al., 2005) have addressed issues related to the general conception of the quarries. Most of them are concerned with the explanation of the quarry reclamation techniques, the impacts generated from quarries, the reuse of quarries' wastes in construction materials, the reclamation of an extraction site in a specific area, and the reuse of the quarry dust on the soil. The current research, on the contrary, attempts to highlight the importance of the quarry rehabilitation typologies describing several made examples in a unique chapter, comparing, and discussing them to investigate, in the next chapters, new strategies of recovery.

3.5 Reuse and Reinvention in the Contemporary Era: Quarry Current Trends

3.5.1 Some Selected Case Studies About the Environmental Potentialities

The discussion of the criteria, the typologies of quarry rehabilitation and the general approaches of the interventions are significant for the comprehension of the conversion of an abandoned extractive site into a renewed and attractive place, with a new utility (Milgrom, 2008). In this section, several examples of quarry reconversion and recovery experiences are schematically illustrated to understand the current tendencies about the topic and to identify useful markers and design approaches. According to recent studies, the most frequent recoveries of dismissed quarries have transformed the industrial environment into leisure parks, public gardens, sports areas, cultural and artistic places, technological sites, and productive spaces for the agriculture.

According to this premise, the present section shows some of the most meaningful interventions and, successively, it deepens the more emblematic cases of study. This program is far from an exhaustive review compared to the globality of the real examples; therefore, it represents a substantial base of the latest trends, in the transformation sector, which does not provide only the naturalistic recovery of the post-extractive sites. Thus, with the chosen cases of study, the research attempts to perform a general screening of contemporary experiences. As indicated in Table 3.1 the selected projects have been classified in conformity with their different uses.

From this general selection, the research explores some peculiarities of the extraction sites, scarcely considered until now, deepening some of the specific projects illustrated in the previous table. The previously mentioned peculiarities concern the concept of the environment, due to the strong imbalance started by humanity against the planet. Some of these singularities are mainly related to the environmental potentialities, which include environmental control (mitigation and adaptation), microclimates, geodiversity, and biodiversity. The term "environment" has ancient Greek and Latin origins, and it is a central element in various disciplines, both in the scientific and humanistic domains. Despite multiple connotations, the concept represents an open system of relations between biotic and abiotic elements that interact in a common context, using perpetual energy exchanges (Di Palma, 2007). Beyond the contextualization of meanings that the term assumes, this climate of changes, sensible to the environmental factors, embraces, albeit belatedly, the architectural, landscape, and urbanistic culture. It is necessary to consider the ecological study in

Table 3.1 Some of the selected quarry rehabilitation projects (the examples with sustainable and environmental components are illustrated in bold-italic) (Katia Talento, 2020)

Naturalistic	Recreational	Cultural	Productive	Facilities
Karsdorf vineyard (Germany) Vineyard and afforestation in limestone quarries	***Brick pit ring*** **(Australia)** Recreational Park promoting biodiversity with presence of rare species	**Montraker (Croatia)** International summer school of sculpture located in an active and Vrasar stone quarry	**Portel winery (France)** Winery and quarry museum in a gypsum quarry	**Beinheim port (France)** River port on the Rhine for barges and boats in a sand and gravel quarry
Rheinelbe (Germany) Naturalistic park with garden school, seat of IBA Emscher Park in a coal mine	**Rordal Lake Park (Denmark)** Park for aquatic sports with open-air theater in a gypsum quarry	**Crazannes (France)** Rest area with open-air museum in a marble quarry	**Druridge Bay (UK)** Territorial Park with lakes, seaside resort and naturalistic oasis in a coal quarry in Cresswell	***Hidrarchy project*** **(Italy)** The quarries became reservoirs to mitigate hydric system variations
Zanderij quarry (Holland) Terraced park with botanical collections in an aggregate quarry	**Bois-le-Roi (France)** Public park with bathing area in a sand quarry	**Westpark (Germany)** Urban park with showrooms, exhibition spaces, greenhouses, and amphitheater in gravel quarries	**Blue water (UK)** Shopping center in a limestone quarry in Greenhithe	**Memory Forest (Sweden)** Cemetery park in Enskede in a gravel quarry
Blue circle (UK) Wooded areas and golf course in clay quarries	**Firminy (France)** Stadium and youth and culture home in a coal quarry designed by Le Corbusier	***Moray Sacred Valley*** **(USA)** Opencast agronomic laboratory	**Denia Mountain (Spain)** Commercial and touristic center in a limestone quarry, with hotel, auditorium, parking, and terraced gardens	**Igualada cemetery (Spain)** Cemetery park and place to rest and reflect in the solitude
Vulcano Croscat **(Spain)** Natural park with geological, botanical, faunistic, and landscape value	**Cergy ponds (France)** Lakes for aquatic sports, swimming pool, tennis, golf courts, equestrian trails, and naturalistic areas in a sand quarry	**Broken circle and** Spiral Hill (Holland) Land art pieces in a sand quarry in Emmen. **Dionyssos quarry (Greece)** Strolling park to valorize the industrial memories	**Ca' Trebbia (Italy)** Agricultural recovery in a gravel quarry near Piacenza	***Chicago deep tunnel*** **(USA)** Biggest quarries in the world transformed into stormwater and linked through channels and tunnels

(continued)

Table 3.1 (continued)

Naturalistic	Recreational	Cultural	Productive	Facilities
La Mortella gardens (Italy) Gardens in a trachyte quarry, spaces for artistic manifestations and concerts	**Emscherbruch (Germany)** Landscaped Park with recreational areas (skating rink and mountain bike) in coal mines	**Fossar de la Pedrera (Spain)** Memorial park in a stone quarry in Barcelona	**Quarries hotel (Italy)** Touristic settlement in white tuff quarries in Favignana	**Spiral Hill (Netherlands)** Sand dock revolving around a rock
Isola Giarola quarry (Italy) Area for recreational, naturalistic, and productive purposes in a sand quarry	**Braga Stadium (Portugal)** Stadium in a stone quarry on the north side of Monte Castro in Braga	**Tindaya Mountain (Spain)** Exhibition space and land art intervention in a marble quarry	**Intercontinental hotel (China)** Touristic building constructed and attached to the vertical wall of the quarry	**Skogskyrkogarden (Sweden)** Cemetery park and masterpiece of recycled places
Luneo quarry (Italy) Oasi WWF in a clay quarry with spontaneous re-naturalization	**Henry Palmisano Park (USA)** Recreational park conceived in a multifunctional and adaptive way. It also serves as a buffer in flooding situations	**Geo-mining quarry (Italy)** Place with interesting geological features in Sardegna	**Hotel in a quarry (China)** Touristic building suspended on the quarry surfaces	
Lignite quarry (Germany) Re-naturalization in the Lusatia region	**Gym free climbing (Italy)** Rock gym in a stone quarry in Rome	**Lìthica (Spain)** Public Park, cultural center with also green spaces	**Quarry botanic garden (China)** Touristic space for visitors approaching natural landscape and experiencing the culture of quarrying industry	
Las Medulas (Spain) Spontaneous re-naturalization in former golden mines	***Atlanta's Westside Park*** **(USA)** Recreational Park water reserve container	**La Palomba Park (Italy)** Sculptural park in Matera archaeological park of rock churches	**Pierre et Vacances costa Plana** (France) Touristic settlement in the French quarries	

(continued)

Table 3.1 (continued)

Naturalistic	Recreational	Cultural	Productive	Facilities
***Zanderij Crailoo* (Holland)** Natural reserve close to the Dutch railway	**St. Margarethen quarry (Austria)** Space with auditorium for open-air concerts	**Royal Botanic Gardens Victoria (Australia)** Educational purposes to show different Australian micro-landscapes		

a broader sense, examining it systemically and on a large space-time scale (Fabbri, 2007). The following provides some potentialities for the transformation of industrial places, concerning specific environmental aspects.

3.5.2 Environmental Control: Mitigation and Adaptation

Current climate changes are increasing the temperature of the Earth's surface and seas and altering the rainfall patterns. This phenomenon produces considerable effects, including rising sea levels, risks of erosion, desertification, and the probable growing strength of natural disasters of meteorological origins. To face the impacts of climate change is important to reduce greenhouse gas emissions, preventing actions of adaptation to its impacts (Verburg et al., 2019). Post-extractive areas can have salient application possibilities, with minimal interventions, to provide answers to future climate challenges, making the territories less vulnerable. This method aims to prevent the risks and the negative consequences to create economic benefits and to decrease the damage to ecosystems, human health, assets, and infrastructures (European Commission, 2017).

The concentration of greenhouse gases in the atmosphere is the principal factor that causes a substantial increase in temperature and rainfall. Different problems can include floods, wildfires, and melting glaciers.

The dismissed quarries often present particular features to adapt them to climate change and to control the risks. For this reason, after their exploitation, quarries can be reused for water storage in the case of fire emergencies, drought, irrigations, or for other agriculture uses, taking advantage of the capacity of the reservoir (Jawecki et al., 2019; Tan et al., 2019). Unfortunately, not all quarries are suitable for this conversion, only those with special conditions: appropriate dimensions and typology of the reservoir, water volume, soil permeability, capacity of the discharge of water on the walls, loss of water from evaporation, the position of the quarry, and its distance from the community.

Among the numerous studies focused on this topic, the work **HIDRARCHY+,** in the Venetian territory (Northern Italy), is particularly sensitive to climate changes.

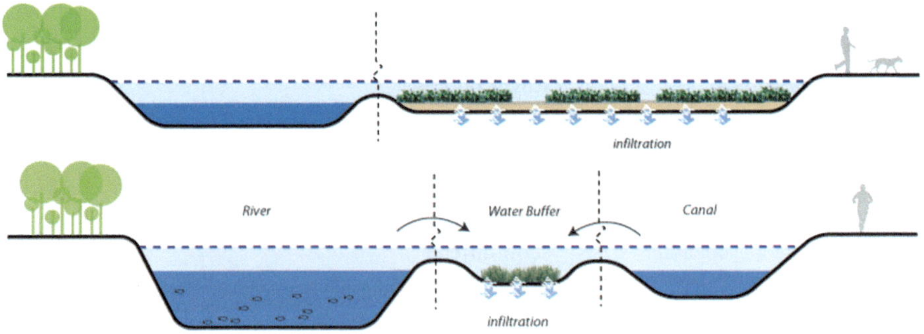

Fig. 3.10 (**a**) Final infiltration by overflowing the full basins into a filtration zone. The excess water is infiltrated before the spring line. © Lorenzo Fabian. (**b**) Flood buffer: canal and river system. © Lorenzo Fabian

The quarries located in this area became reservoirs to mitigate the variations of the hydric system. On the one hand, they are storage areas for the dry season and the drought periods, and on the other hand, they represent the buffing and infiltration structures against flooding (Fig. 3.10a, b) (Fabian & Viganò, 2010). The realized recovery interventions in this perspective were rather rare until now.

About flood problems, there are two significant projects made in the Chicago area: **Henry Palmisano Park** and **Chicago's Calumet Deep Tunnel** in the Thornton Quarry. The first one is a recreational park conceived in a multifunctional and adaptive manner; this park also serves as a buffer in flooding situations (Mortice, 2015). The space can contain 5.56 million gallons of water using bioswales, wetland cells, and a retention pond, and it saves 10.5 million gallons of potable water by using native prairie plants, which requires no irrigation (Fig. 3.11).

The second work was conceived within one of the biggest quarries in the world, transformed into a stormwater detention structure. It is linked, through channels systems and tunnels, to the south metropolitan area of Chicago and the suburb of Cook County (Fig. 3.12). Apart from reducing stormwater flooding and damage and removing occasional sewage infested overflows from Chicago's streets, the Thornton Quarry project is also for the benefit of much of Chicago's remaining waterways, the city's stream, rivers, and lakes, through diverting sewage-tainted water. In a city in which the Illinois Environment Protection Agency believes sewage overflow to be the "water quality issue of today," this is a massive boost (Neil, 2015). The quarry will be the largest control reservoir to prevent floods, with an estimated capacity of 7.9 million gallons.

Finally, **Atlanta's Westside Park**, in the former Bellwood Quarry, can also be cited. This project represents one of the greatest examples (still in progress) of a water reserve container to avoid drought, and it is part of the new program of hydric supply for Atlanta city. The pit will also be part of integrating a large recreational park (Fig. 3.13).

Fig. 3.11 Henry Palmisano Park. © Site_design_group_ltd

Fig. 3.12 Chicago's Calumet Deep Tunnel. © John Trautschold

Fig. 3.13 Atlanta's Westside Park. © Eric Mandel/City of Atlanta/Atlanta Business Chronicle

3.5.3 Microclimate

The extraction sites normally constitute components of discontinuity in landscape. This disconnection is the consequence of multiple factors, from morphological to vegetational. The physical elements considerably affect the microclimate conditions of the examined area and its surroundings. For instance, rock landscapes, without vegetation, provoke a conspicuous enhancement of the temperature, and a terraced system significantly increases the sunny surface exposure of a quarry. As aforementioned, the most relevant features of a place are related not only to the shape of the ground but also to the superficial structure, its composition, color, and the presence or absence of flora and water. These critical characteristics influence the local meteorological processes and, consequently, the vocation of the transformation of the area to the possibilities connected to agriculture fields or the creation of habitats and shelters for native species.

At this point, it appears pertinent to refer to the case of the ancient archaeological site known as "**the sacred valley of Moray**" in Peru, where unusual Inca ruins exist on an upland of 3500 m of altitude (Fig. 3.14). They morphologically arise as huge circular terraced depressions, in which the space between them has a depth of 30 m. The more probable use of these depressions was that of quarries and, successively, a massive opencast agronomic laboratory, due to the singular features of the site. In this respect, their orientation relative to the sun and wind generates a consistent disparity of temperature of about 15 degrees C° from the upper to the lower part. This vast thermal variation was used by the Inca to naturally investigate the effects

Fig. 3.14 Sacred Valley of Moray. (NewAdventuresPeru.com, 2020)

of diverse climate conditions on their cultivation (Aliberti, 2015). This notable open-air laboratory offers a glimpse of the science developed by this civilization, which has now disappeared (Angelini, 2019).

3.5.4 Biodiversity and Geodiversity

Based on the previous considerations, the research found that the places altered by natural disturbances presented great potential from the ecological point of view, and they also represented real "containers" of rare biodiversity (Sayer et al., 2017). Many geological faults become ecological corridors and spaces of connectivity, shelter, and biological transit, ensuring the survival of particular biomes. This finding suggests that the modified areas, through the extraction activity, can generate other repositories of biodiversity. If adequately planned, the open-cast quarries can contribute to the creation of new habitats for flora and wildlife (Gagné et al., 2015).

Regarding this specific vision, numerous environmental studies conducted in France and Germany have demonstrated that some rare protected species find refuge in the ancient extraction sites. Part of the cited study, concerning 35 French stone quarries, highlights the presence of new species within these former industrial pits, such as birds, amphibians, and reptiles, including rare and endangered species. Thus, if there are certain context conditions, the protection of biodiversity can be considered meaningful (Aliberti, 2015). Two significant projects that promote this conception are the **Brick Pit Ring** in Australia and **Zanderij Crailoo** in Holland.

The first one is the last material evidence of a vast working industry at Homebush Bay, Sydney. It is archetypal and primitive, raw, deprived, and transformed. The Brick Pit is first a place of exceptional human effort, now arrested, and it is a portrait of the altered land through use. Equally, this is a place of adaptation where new sustainable technologies replaced a leftover industry, as well as a refuge for rare and endangered species, such as the green and golden bell frog. Durbach Block architects have valorized this place conceiving a large recreational aerial walkway with a circular shape, which serves for outdoor exhibitions and as an observatory of the mentioned species.

The Ring Walk, located 20 m above the brick pit floor, gives the area an original urban connection and presence within the Sydney Olympic Park. A simple ordering structure, a circular walkway that facilitates access to the pit, recognizing the fragility of the environment and its pure form. The consistent level of the ring also expresses the different depths of the excavation. The Ring Walk allows for both a 10-min walk and a longer layered experience, through wide and shaded sections of the platform. The exterior surface is a colorful screen, an exhibition of mesh and glass panels. The ring, thanks to the presence of an audio guide, provides visitors with different perspectives on the history of the place and its use as a wildlife refuge. The steel structure is a slender and delicate intervention within the massive roughness of the pit, and it appears to tiptoe across this fragile spot (Fig. 3.15) (De Poli & Incerti, 2014).

Zanderij Crailoo is today a natural reserve developed close to the Dutch railway. This area of 65 hectares originally represented the perfect place to extract sand. After its exploitation, the industrial site was converted into a multifunctional area having an ecological overpass next to the railroad. This green corridor, 800 m long and 50 m wide, provides safe passage of the wildlife from side to side of the reserve, bisected by major roadways (Fig. 3.16). Rather than try their luck with the traffic,

Fig. 3.15 Brick Pit Ring. © Peter Hyatt

Fig. 3.16 Zanderij Crailoo. © Province of Noord-Holland/Theo Baart

native populations of deer, wild boar, and the endangered European badger can wander across on one of these designated wildlife bridges that were planted to simulate the surrounding environments (Zeunert, 2017). Therefore, structures like Natuurbrug Zanderij Crailoo are exciting examples of the union between conservation efforts and architectural innovation (Richards, 2018).

The concept of biodiversity, jointly with the notion of geodiversity, forms the natural diversity of a precise area. The term "geodiversity" describes the range of geological, geomorphological, pedological, and hydrological features of a place (Gray, 2018). The extraction activity is closely linked to the physical characteristics of the minerals from mining. Quarries provide many of the most valuable geological exposures and are, therefore, a vital resource for geological education, training, and research. The use of this resource, at the discretion of quarry operators, may range from the involvement of schools and local communities in discovering and understanding their local, natural heritage, to more detailed scientific research and the conservation or recording of relevant geological evidence (Thompson et al., 2018).

In this respect, excavation reveals sedimentary layers that would otherwise be unnoticed. Excavation can also show rare situations, attractive for scientific, educational, and scenographic value. The various elements, in some cases, can present peculiar geomorphological qualities, such as shape, color, process of sedimentation, geological discordances, layer relations, and fossils. Considerable dinosaur tracks, for instance, have been discovered in a quarry in South Tytherington, England. For this purpose, the geomorphological components heavily affected the definition of an extractive landscape and, as such, they are part of the local heritage. A remarkable number of projects have recently persevered in the objective of valorizing and preserving the geodiversity peculiarities. Several of these projects only provided actions of protection of the relevant parts, improving safety and accessibility and establishing educational and informational paths dedicated to geodiversity.

Volcano Croscat, in Garrotxa (Spain), is one of many examples that singularly interpret this notion. The natural park of La Garrotxa Volcanic Zone is an area of magnificent geological, botanical, faunistic, and landscape value and, despite the presence of human activity, its natural environment has been well preserved. This protected area includes about 50 stored volcanic cones, which manifests a wide variety of Strombolian and phreatomagmatic volcanic outputs. The volcanic activity, combined with the humid climate aspects of this space, gave rise to fertile soils and, consequently, a vast range of biodiversity. Since its declaration as a natural park in 1982, it has become a famous attraction for local and foreign tourism.

The recovery methodology of the quarry overthrows the negative thought of "hurt," on the contrary, it has emphasized the potentialities instead of being "healed." In this way, the enormous cut in the hill, made by the extractive industry, has been conceived as an educational activity, thanks to its geological exceptionality. This damage permits the observation of the interior geological structure of the volcanic cone, as if it is a section of a drawing (Fig. 3.17). La Garrotxa Volcanic Zone is a fitting case of study to investigate the impact and sustainability of geo-tourism on preserved volcanic land, as it has allowed the transmutation of a poorly known

Fig. 3.17 Vulcano Croscat. © Oscar Rodbag/Turisme_Garrotxa

territory into one of the most famous and most visited geo-sites of the Catalonia community.

The protection of this volcanic area symbolized the end of quarrying actions that significantly destroyed most of its volcanoes, but also produced a chance to increase the zone for tourism. Geo-tourism is a growing business in our society, a sustainable type of tourism based on an interdisciplinary combination of the tourism between industry and the maintenance of the geological heritage. In parallel, geo-tourism also improves the economic and social expansion of local communities. La Garrotxa Volcanic Park proved that using geo-conservation as a means of sustainable tourism is a beneficial strategy. If well managed, this strategy can bring economic and social gains to the area, not only in tourism but also in the global image of the territory (Planaguma & Martí, 2018).

3.6 Chapter Resume

This chapter has focused on human activities that have strongly influenced our environment, and it has discussed the extraction industry, which mirrors the activity that has caused, and still causes, changes in the land. As it has been shown in this chapter, after their exploitation, it is important to transform industrial and abandoned places into something with architectural and environmental value. Numerous notable typologies of recovery exist, but in this chapter, the research has mainly promoted those belonging to the architectural and landscape domains, analyzing the subsequent environmental implications. Through regeneration strategies, the previously devastated industrial territory acquires a new aspect, function, and meaning.

Quarries' spaces become the location of various architectonic scenarios, and, according to the typology of the excavation and its main features, they incorporate a wide range of structures and uses. Quarries have been transformed into large public parks, botanic gardens, natural areas of considerable environmental added value, auditoriums, tourism buildings, infrastructures, shopping malls, cemeteries, and many other leisure areas. In addition to conferring a new life to the degraded territory, this aspect affects the economy of the place where the quarries are located, reducing the need for new land.

Most of the time, the extraction of material from quarries involves discarding a considerable amount of it. This quantity is abandoned at the side of the pit and accumulated in piles of rejected material. Due to their vast dimensions, the waste heaps are considered destructive elements of the landscape that cannot be demolished or stored elsewhere due to economic factors. Not including these accumulations of waste in the process of regeneration of a quarry represent a significant disadvantage. On the contrary, they could be integrated into any typology of recovery to complete the process and respect the theory of the circular economy.

In this regard, it is necessary to underline that this literature review is only a minor part of the broader research focused on innovative strategies for quarry rehabilitation. They suggest replacing not only the former industrial pits but also the giant heaps of remains located at their side, linking the two components (Talento et al., 2019a). As it will be described in the next chapters, this study is also concerned with the Anticlinal of Estremoz, Portugal, consisting of an area of 280 km^2 formed by approximately 185 marble quarries, each of which is flanked by an enormous pile of waste (that can reach up to 40 m in height). These enormous mounds of remains symbolize a different meaning to requalify the degraded landscape and to reuse the wastes in situ (Talento et al., 2018). They are the main factors to promote tourism in the place, through the replanning of the abandoned extractive area and the addition of new structures built inside the waste heaps. These interventions will be connected to the inactive quarries, and they will be incorporated into the surrounding landscape.

Based on these innovations, the current review, through the conception of the schematization related to the different rehabilitation typologies of a quarry and their discussions (previously illustrated), only constitutes the starting point on which to add new methodologies of rehabilitation, never before experienced. In this way, the research aspires to produce new issues, in both the scientific and practical sectors, related to the inclusion of waste heaps in the recovery interventions of a quarry.

Bibliography

Aliberti, L. (2015). Preadattamento e trasformabilità dei siti alterati da attività estrattiva. Università degli Studi di Napoli Federico II Scuola Politecnica e delle Scienze di Base Dipartimento di Architettura, Dottorato in Progettazione architettonica e tecnologie innovative per la sostenibilità ambientale XXVII ciclo.

Angelini. (2019). *Mexico dream* (ebook).

Architeam. (2013, January). Braga Municipal Stadium. http://www.architravel.com/architravel/building/braga-municipal-stadium/

Archzine, A. (2014, November). Eduardo Souto Moura: Edifício do Aproveitamento Hidroelétrico, Foz Tua. http://afasiaarchzine.com/2014/11/eduardo-souto-moura/

Bellini, O. E. (2013). Grey to green: Il verde tecnologico come risposta ambientale per le infrastrutture urbane. TECHNE 05.

Blanchette, M. L., & Lund, M. A. (2016). Pit lakes are a global legacy of mining: An integrated approach to achieving sustainable ecosystems and value for communities. *Current Opinion in Environmental Sustainability, 23*, 28–34.

Bull, C. (2006). The Australian garden at the Royal Botanic Gardens Cranbourne, in Victoria uses abstractions of iconic Australian landscapes. *Landscape Architecture Australia, 111*, 38–44.

Cesarin, C., & Fornari, D. (2010). *Estetiche del camouflage* (Et al).

Chetia, M., & Sridharan, A. (2016). A review on the influence of rock quarry dust on geotechnical properties of soil. *Asce Library*. https://doi.org/10.1061/9780784480151.019

Cifani, G. (2008). Architettura romana arcaica: Edilizia e società tra monarchia e repubblica. L'ERMA di BRETSCHNEIDER.

Conan, M. (2012). The quarries of Crazannes: Bernard Lassus's landscape approach to cultural diversity. *Studies in the History of Gardens & Designed Landscapes An International Quarterly, 23*(2003), 347–365. https://doi.org/10.1080/14601176.2003.10435307

Cook, P. (2016). *Architecture workbook: Design through motive*. Wiley.

Crutzen, P. J. (2005). *Benvenuti nell'Antropocene. L'uomo ha cambiato il clima, la Terra entra in una nuova era*. Mondadori.

D'Amato, C. (2005). *Paesaggi di cava. L'attività estrattiva e il paesaggio in area mediterranea*. Uniongrafica Corcelli Editrice.
Damigos, D., & Kaliampakos, D. (2003). Assessing the benefits of reclaiming urban quarries: A CVM analysis. *Landscape and Urban Planning, 64*, 249–258. https://doi.org/10.1016/S0169-2046(02)00243-8
Dávid, L., & Patrick, C. (1998). Quarrying as an anthropogenic geomorphological activity. In *Anthropogenic aspects of geographical environment transformations*.
De Poli, M., & Incerti, G. (2014). *Atlante dei paesaggi riciclati*. Skira editore.
de Wit, S. (2016). Sensory landscape experience: Stepping outside the visual landscape of the motorway in the garden of birds. *Journal of Landscape Architecture, 11*(3), 20–31. https://doi.org/10.1080/18626033.2016.1252157
Dell'Aira, P. V., Grimaldi, A., Guarini, P., & Lambertucci, F. (2015). Sottosuoli urbani: La progettazione della città che scende. *Quodlibet.*
Deshaies, M. (2018). The new post-mining energy landscapes in the lignite basin of lower Lusatia (Germany). *Europa Regional, 25.2017*(3-4), 29–41.
Di Palma, E. A. (2007). *Sistema ambiente e sistemi informativi, la rappresentazione della complessità nella progettazione ambientale*. Università "G. D'Annunzio".
Diord, S., Magalhães, F., Cunha, Á., Caetano, E., & Martins, N. (2017). Automated modal tracking in a football stadium suspension roof for detection of structural changes. *Structural Control and Health Monitoring, 24*. https://doi.org/10.1002/stc.2006
Eldredge, N., & Pievani, T. (2010). AA.VV. Ecosphera. 5, 22.
European Commission. (2017). *White paper on the future of Europe*. https://doi.org/10.2775/32364
Fabbri, P. (2007). *Principi ecologici per la progettazione del paesaggio* (Franco Angeli).
Fabian, L., & Viganò, P. (2010). *Extreme city. Climate change and the transformation of the waterscape*. IUAV University.
Fallacara, G., Colella, M., & Barberio, M. (2016). Architecture and didactics: A new cultural centre for Matera, European capital of culture 2019. *International Journal of Contemporary Architecture "The New ARCH"*, 3(2). https://doi.org/10.14621/tna.20160204
Fidalgo, P. (2017). *Estudos de paisagem* (Vol. 3). Instituto de História Contemporânea—NOVA FCSH.
Fuentes, A., & Baynes-Rock, M. (2017). Anthropogenic landscapes, human action and the process of co-construction with other species: Making Anthromes in the Anthropocene. *Land, 6(1*, 15. https://doi.org/10.3390/land6010015
Gagné, S. A., Eigenbrod, F., Bert, D. G., Cunnington, G. M., Olson, L. T., Smith, A. C., & Fahrig, L. (2015). A simple landscape design framework for biodiversity conservation. *Landscape and Urban Planning, 136*, 13–27. https://doi.org/10.1016/j.landurbplan.2014.11.006
Galetakis, M., & Soultana, A. (2016). A review on the utilization of quarry and ornamental stone industry fine by-products in the construction sector. *Construction and Building Materials, 102*(15), 769–781. https://doi.org/10.1016/j.conbuildmat.2015.10.204
Goudie, A. (2018). The human impact in geomorphology—50 years of change. *Geomorphology, 106601*. https://doi.org/10.1016/j.geomorph.2018.12.002
Gray, M. (2018). Chapter 1—Geodiversity: The backbone of geoheritage and geoconservation. *Geoheritage Assessment, Protection, and Management*, 13–25. https://doi.org/10.1016/B978-0-12-809531-7.00001-0
Guallart Architects. (n.d.). *Denia mountain*. Retrieved 13 December 2019, from http://www.guallart.com/projects/denia-mountain
Holden, R. (2003). *New landscape design*. Architectural press.
Jawecki, B., Kowalczyk, T., & Feng, Y. (2019). The evaluation of the possibility to use the water from Quarry Lakes for irrigation. *Journal of Ecological Engineering, 20*, 188. https://doi.org/10.12911/22998993/112490
Kirtan, P., & Siddharth, S. (2015). State of the art review on utilization of quarry waste replacing fine aggregate in concrete. *International Journal Of Research in Engineering and Technology, 04*.

Kus, J., Dolezych, M., Schneider, W., Hofmann, T., & Visiné Rajczi, E. (2019). Coal petrological and xylotomical characterization of Miocene lignites and in-situ fossil tree stumps and trunks from Lusatia region, Germany: Palaeoenvironment and taphonomy assessment. *International Journal of Coal Geology., 217*, 103283. https://doi.org/10.1016/j.coal.2019.103283

Langer, W. H. (2001). Potential environmental impacts of quarrying stone in karst- A literature review. *USGS Numbered Series, 484*. https://doi.org/10.3133/ofr01484

Langer, P. (2019). The Muskau bend: Mining space from different perspectives. *IOP Conference Series: Materials Science and Engineering, 471*(Session 10).

Legwaila, I. A., Lange, E., & Cripps, J. (2015). Quarry reclamation in England: A review of techniques. *JASMR, 4*, 55. https://doi.org/10.21000/JASMR15020055

Lovelock, J. (2006). *The revenge of Gaia: Why the earth is fighting Back—And how we can still save humanity*. Allen Lane.

Magnaghi, A. (2010). Il progetto locale. *Verso la coscienza di luogo* (Bollati Boringhieri).

Marcantonio, R., & Fuentes, A. (2020). A clear past and a murky future: Life in the Anthropocene on the Pampana River, Sierra Leone. *Land, 9*(3), 72.

Maza, R. (2004). Estadio de Braga. *TC Cuadernos: Eduardo Souto Moura Obra Reciente, 12*, 198–227.

[MEA] Millennium Ecosystem Assessment. (2005). *Ecosystems and human well-being: Synthesis*. Island Press.

Milgrom, T. (2008). Environmental aspects of rehabilitating abandoned quarries: Israel as a case study. *Landscape and Urban Planning, 87*(3), 172–179. https://doi.org/10.1016/j.landurbplan.2008.06.007

Montañez-Escalante, P., García-Barrios, L., & Jiménez-Osornio, J. J. (2005). Quarry reclamation in Mérida, Yucatán, México: A review on achievements and current limitations. *Tropical and Subtropical Agroecosystems, 5*(3), 101–108.

Mortice, Z. (2015). The history of Chicago is buried under one of its least-known and most astonishing parks [Palmisano Park]. *Landscape Architecture, 105*(11), 150–163.

Morton, T., & Müller, K. (2016). Lusatia and the coal conundrum: The lived experience of the German Energiewende. *Energy Policy, 99*, 277. https://doi.org/10.1016/j.enpol.2016.05.024

Mouflis, G. D., Gitas, I. Z., Iliadou, S., & Mitri, G. H. (2008). Assessment of the visual impact of marble quarry expansion (1984–2000) on the landscape of Thasos Island, NE Greece. *Landscape and Urban Planning, 86*, 92–102. https://doi.org/10.1016/j.landurbplan.2007.12.009

Neil, K. (2015, September). Chicago's South Side Suburbs to Remain Dry. The Colonial Capital Blog – Window on Chicago.

Planaguma, L., & Martí, J. (2018). Geotourism at the Natural Park of La Garrotxa volcanic zone (Catalonia, Spain): Impact, viability, and sustainability. *Geosciences, 8*. https://doi.org/10.3390/geosciences8080295

Preite, M. (2017). *Paesaggi industriali e patrimonio Unesco*. C&P Adver Effigi.

Richards, M. A. (2018). *Regreening the built environment: Nature, green space, and sustainability*. Routledge.

Rosiere, R. E., Engle, D. M., & Cadle, J. M. (1989). Revegetation of Tripoli quarries in the Ozark highlands of Oklahoma. *Landscape and Urban Planning, 17*, 175–188. https://doi.org/10.1016/0169-2046(89)90025-X

Rybnikovaa, L. S., Rybnikov, P. A., & Tarasovab, I. V. (2017). Geoecological challenges of mined-put open pit area use in the Ural. *Journal of Mining Science, 53*(1), 181–190.

Salema, I. (2005, January). *Transformar uma geografia é a ambição de todos os arquitectos*. Publico. https://www.publico.pt/2005/01/27/jornal/transformar-uma-geografia-e-a-ambicao-de-todos-os-arquitectos-2562

Sayer, J., Margules, C., & Klintuni Boedhihartono, A. (2017). Will biodiversity be conserved in locally-managed forests? *Land, 6(1*, 6. https://doi.org/10.3390/land6010006

Shi, Z., Wu, Y., Chiu, Y., Wu, F., & Shi, C. (2020). Dynamic linkages among mining production and land rehabilitation efficiency in China. *Land, 9*(3), 76. https://doi.org/10.3390/land9030076

Shlomo Aronson Architects. (2020, January). *Shlomo Aronson Architects site* [Official site]. https://www.s-aronson.co.il/

Souza, E. (2013, May). *AD classics: Igualada Cemetery/Enric Miralles + Estudio Carme Pinos*. Archdaily. https://www.archdaily.com/375034/ad-classics-igualada-cemetery-enric-miralles-carme-pinos.

Talento, K., Miguel, A., & Kullberg, J. C. (2018). The metamorphosis of the landscape: The adaptive reuse of marble waste. In *The 5th international conference on architecture and built environment with AWARDs S.ARCH 2018 Conference proceeding*. Get It Published.

Talento, K., Amado, M., & Kullberg, J. C. (2019a). Eco-friendly reuse of marble wastes in landscape and architecture. *The Sustainable City XIII. WIT Transactions on Ecology and the Environment, 238*, 665–677. https://doi.org/10.2495/SC190571

Talento, K., Amado, M., & Kullberg, J. C. (2019b). Landscape—A review with a European perspective. *Land, 8*(85). https://doi.org/10.3390/land8060085

Tan, F., Jiao, Y.-Y., Wang, H., Liu, Y., Tian, H., & Cheng, Y. (2019). Reclamation and reuse of abandoned quarry: A case study of ice World & Water Park in Changsha. *Tunnelling and Underground Space Technology, 85*, 259–267. https://doi.org/10.1016/j.tust.2018.12.009

Tang, Y., Jia, H.-F., Jiang, Q.-G., & Wang. (2011). Comprehensive rehabilitation planning of deserted pits and the case study in plain area of Beijing, China. *Landscape and Urban Planning, 99*, 123–132. https://doi.org/10.1016/j.landurbplan.2010.09.003

Tarolli, P., Cao, W., Sofia, G., Evans, D., & Ellis, E. C. (2019). From features to fingerprints: A general diagnostic framework for anthropogenic geomorphology. *Progress in Physical Geography: Earth and Environment, 43*(1), 95–128. https://doi.org/10.1177/0309133318825284

Thompson, A., Poole, J., & Carroll, L. (2018). Geodiversity action plans for aggregates companies. *Proceedings of the 14th Extractive Industry Geology Conference*, 78–82.

Tsolaki-Fiaka, S., Bathrellos, G. D., & Skilodimou, H. D. (2018). Multi-criteria decision analysis for an abandoned quarry in the Evros region (NE Greece). *Land, MDPI, 7*, 43. https://doi.org/10.3390/land7020043

Verburg, R., Rahn, E., Verweij, P., Van Kuijk, M., & Ghazoul, J. (2019). An innovation perspective to climate change adaptation in coffee systems. *Environmental Science & Policy, 97*, 16–24. https://doi.org/10.1016/j.envsci.2019.03.017

Zeunert, J. (2017). *Landscape architecture and environmental sustainability: Creating positive change through design*. Bloomsbury.

Chapter 4
Waste Reuse from Quarries

Rebuilding instead of building: building on, around, inside, with waste materials.
To live in ruins instead of building; re-naturalize rather than re-urbanize.

Pippo Ciorra

4.1 Introduction

From the artistic concept of ready-made to circular economy, defining innovative typologies of reuse for waste heaps from quarries and mines.

In 1913, the artist Marcel Duchamp presented to the public his first *ready-made*, "*Roue de bicyclette*." The term refers to an everyday object that arises to a piece of art thanks to the ability of the artist to remove its original use, by exalting it through an operation of choice, casual individuation, or isolation. After more than one hundred years, that operation still preserves a subversive foundation, and it represented a milestone for the definition of the aesthetic parameters of the twentieth century. Picasso already brought the *"objects trouvés"* (found objects, recovered with minimal interventions, and presented as works of art) into art galleries, but contrary to the *ready-made*, they did not lose their original meaning. In reference to his work "Bull Head," a handlebar, however, evocative of Taurus horns, was still a component of a bicycle frame. Duchamp made a waste instead. His bicycle's reversed wheel, inserted above a stool, and deprived of its tire, represented a self-sufficient work. It could be admired from the top of a pedestal as a magnificent piece of art. The wheel, decontextualized from its frame, acquired a new meaning, and it became, from an everyday object, an artistic *unicum*. This change of destination of use introduced an innovative notion of art since it promoted values scarcely verified until that moment. Thanks to the conception of the ready-made, the selection prevailed over the invention, and beauty corresponded to the essence of an object, undermining the concept

K. Talento, *Sustainable Buildings from Marble Waste Heaps*,
https://doi.org/10.1007/978-3-031-72146-5_4

of form. For this reason, in the twentieth century, the poetics of transformation followed that of creation in the artistic sector (De Poli & Incerti, 2014).

Referring to the architectural field, this understanding was unsuitable for the coeval architecture where the idea of beauty exclusively corresponded to the "new architecture." In this way, there were no connections with the past and the identity of the place. Nowadays, on the contrary, the preexisting elements influence the concept of beauty and the recovery of many places that can obtain new meanings. These recovery projects enhance the importance of the metamorphosis as a symbol of human continuity that reveals the existent. As happens in the artistic sector with the ideation of the ready-made, the architectural rehabilitation process could not include the operation of "erasing" the traces of the past, but it should incorporate the action of protecting the best cases of architectures and landscapes, as testimonies of the history of the place (Talento et al., 2019). Industrial territories are the most representative examples of areas that need rehabilitation processes to preserve their identity. As mentioned in the previous chapter, several worldwide projects exist to demonstrate the necessity of rehabilitation operation (Baillie, 1992; Dal Sasso et al., 2012; El-Fadel et al., 2001; Talento et al., 2020).

Countless man-changed territories represent the motivation of providing a different use to what has been exploited and abandoned. More often, territory changes consist both in the alteration of the place morphology and in the presence of many wastes. Nowadays, the rehabilitation of quarries is a common practice; the aspect less prevalent is to include the wastes in the process of restoration. Because of the industrial activity, the wastes, accumulated in huge piles, further contribute to the degradation of the sites. They appear to be disturbing factors, not communicating with the place where they belong, and isolated elements deprived of any use. For this reason, one wrongly excludes the wastes from the process of requalification, preferring to leave them anonymous in the territory, or exporting part of them outside their original place. If one applies the artistic concept to the rehabilitation of quarries, the significant number of wastes could also be considered as an "*object trouvés*" or a "ready-made" and arise into a "piece of art" with a renovated function.

Based on these premises, what happens to the huge accumulations of waste?

After discussing the theory of waste and circular economy, this chapter provides a critical survey about some practical examples of reuse, applied to the dumped material in the landscape and architectural fields. This "mix" contributes to the creation of a new systematic architectural methodology, presented in the fifth chapter of the research, that identifies and organizes the main examples of waste heap reuse to formulate new criteria for their reconversion, filling an existing gap in both architectural and theoretical domains. To the best of the author's knowledge, there is not any similar schematization in the literature, and the idea of dealing with the mounds of waste in the architectural and landscape sectors has been slightly approached until now.

Could waste become an integral part of the architectural design philosophy promoting reuse interventions in the local?

This interrogation represents the key question addressed by the current chapter that promotes the strategy of transforming inert matter into a valuable resource envisaging new life cycles. The work, described in this chapter, examines this option by introducing alternative approaches to how architecture can be contemporary without falling into the trap of linear thinking, based on exclusion and separation. The research describes numerous examples of waste reuse in architecture, which considers the discarded material not as a burden but as a renewable opportunity.

Section 4.2. addresses the concept of waste and circular economy, and it describes the practice of reuse in a discarded context. Section 4.3. analyzes the piles of waste material in the architectural, landscape, and arts fields. In this section, the research selects significant examples of projects conceived with waste material, by differentiating the off-site applications from those in situ and proposing an innovative matrix with distinct hypotheses of reuse. Section 4.4. elaborates a systematic architectural methodology to deal with the reuse of waste heaps. Finally, Sec. 4.5. draws the chapter conclusions.

4.2 Waste Concept and Circular Economy

Recognizing waste as a chance for the architectural production process, it appears essential to reconsider its definition, introducing the conceptual knowledge of it. The specific meaning of waste is apparently irrelevant but extremely crucial; the notion of waste is far from being obvious, and it has been described by several definitions (listed in Table 4.1). The common factor is that waste has been recognized as an unwanted or undesired material, which can include fabricated or natural matter (Hebel et al., 2014). The waste is also something that the possessor has disposed or discarded. Generally, both terms mean "abandonment," but perhaps "disposal" refers to inserting it into a proper place, while "discard" has the connotation of being futile or rejected, set aside (Pongrácz & Pohjola, 2004). The European Directive (European Commission, 2008) prefers to adopt the expression "discard" to expand its reach, including the broadest possible acts of abandonment of things. According to the current European definition, the rejected material is considered more as a potential pollutant than as a possible raw substance (Cheyne & Purdue, 1995).

From the earliest times, human actions have necessarily produced waste. The Mayan Indians in North America accumulated many types of wastes, arranged in a monthly cadence. Whenever needed, they burned or covered the solid waste with a layer of earth, which resulted in the continuous growth of the ground levels in their villages. Until a few decades ago, wastes, instead of being considered as virtues and resources, embodied a dead-end scenario in which they could be "buried from view" as an amorphous matter, without value, to be hidden. Modernization represents the most prolific and uncontrolled line of production of scraps and waste material; it is a collateral effect of economic progress (Hebel et al., 2014).

Being the main topic of this research, the extractive process designs something disruptive in the waste context. Due to the uncertainty of the industrial activity that, at any time, can terminate producing goods and gains, the extractive process

Table 4.1 Selected definitions of waste. (Katia Talento, 2020)

Reference	Definition
European Commission (2008)	Waste shall mean any substance or object, which the holder discards or is required to discard.
OECD (1998)	Wastes are materials other than radioactive materials intended for disposal.
Oxford English dictionary online (n.d.)	Refused matter; unserviceable material remaining over from any process of manufacture; the useless by-products of any industrial process; material or manufactured articles so damaged as to be useless or unsaleable.
UNEP (1989)	Wastes are substances or objects, which are disposed of, are intended to be disposed of, or are required to be disposed of by the provisions of national law.
World Health Organization Expert Committee (1965)	Something, which the owner no longer wants at a given time and space and which has no current or perceived market value.
Wilson (2015)	Unwanted or discarded materials, rejected as useless, unneeded, or surplus to requirements.
Gourlay (1992)	What we do not want or what we fail to use.
Douglas (1966)	Matter out of place.
Steffen (n.d.)	There is no such thing as garbage, just useful stuff in the wrong place.
Freilla (2008)	Waste is what you call something when you have no idea what to do with it. The fact that waste exists anywhere is more a testament to our lack of imagination. If you have a purpose for it, it is no longer waste.
Zero waste America (n.d.)	A waste is a resource that is not safely recycled back into the environment or the marketplace.

represents the image of humanity's discontinuity. The "new" cannot be created without something wasted or destroyed. The extraction of minerals assumes that the birth of the "new" requires the death of the "old." In this case, every new creation could share its destiny with the abandoned elements to conceive a production even more modern. The extraction activity proceeds through tipping points, and it embodies a one-way movement, irreversible and unalterable. The chronicle of this type of industry is a cemetery of exploited, repudiated, and abandoned pits; it is unconceivable without waste. At the height of the Renaissance age, Michelangelo proclaimed the precept that led to modern art. He obtained great harmony in his sculptures using the marble block and removing all unnecessary material. The vision of a perfect form, hidden inside the block of rough stone, precedes his creative act. According to this concept, the waste is the shell that covers that form. It must be denuded of its shell to reveal the features of beauty and perfection. This method transforms waste into an indispensable ingredient of the creative process, giving it the power of transmuting a lowly matter into a valuable one. As the architect Jeanne Gang said, *"the materials destined for landfill have begun to explore an architecture originated in the material itself, rather than in a formal language or design concept"* (World Health Organization Expert Committee, 1965).

Since they are the source of the creation and the most fearsome obstacle to it, wastes are simultaneously divine and sinister. Therefore, wastes are sublime, a mix of attraction and repulsion that arouses admiration and fear (Bauman, 2004). They are also valued as gifts, offered by society to itself (Angélil & Siress, 2010).

According to this citation, the wastes are a gift to release from its "pejorative stigma." They are part of the wealth of society and investments to return. Talking about wastes means equipping a specific sensibility. It expresses the inclination to recognize and accept the qualities of existent materials to transform them, rethinking and reshaping what surrounds us. Reuse is an indispensable component to comprehend how to reprocess an obsolete matter into something that is put back into cycles of use (Wong, 1986). In other words, the process of reusing materials introduces unique research methods, which appear different from those that often concentrate on beginning from scratch (Baker-Brown, 2017). It is possible to resume these considerations under the head of *Tabula Plena.*

On the contrary to *Tabula rasa,* which indicates an operation of design starting with a "blank page," *Tabula Plena* refers to working with what is already in situ (existing materials and buildings), ready for reconfiguration (Bryony, 2016). This innovative formulation represents the challenge of the twenty-first century to improve sensibilities around us, reconfiguring the existent to create new places and environments.

The word "reuse" and, consequently, the regeneration of fabricated elements is also at the core of the theory of the circular economy. This concept often suggests a way to reduce wastes (Chung & Poon, 2000) and to protect the environment, by continuously using and reusing resources. The circular economy aims to maximize the circularity of matters in industrial society, by designing products and buildings that can be dismantled, repaired, and reused, bypassing the creation of wastes (Gorgolewski, 2018; House of Commons Environmental Audit Committee, 2014; Rhys, 2011; Till & Schneider, 2012). It implies a variation of the use, location, or ownership, and it is a concept defined by the imperative to approach climate change, which entails decreasing carbon emissions from human activities (Rhys, 2011). To better understand the concept, one can cite the project of Bailey bridges (Baker-Brown, 2017), representative cases of military reinvention. Today, these are reusable structures for temporary applications such as tennis halls, sports events, funfairs, and store cars in winter. The notion of the circular economy has taken part in human growth since the beginning as an approach to surmount poverty and scarcity of resources. In this regard, the theorist Donella Meadows suggests that scarcity happens when resource streams are forced or exhausted (Gorgolewski, 2018). In numerous situations, it is connected to the availability, energy power of processing, cost of extraction, processing, and transport. As suggested by Till and Schneider (Till & Schneider, 2012), scarcity, in the architectural context, is more than a lack of supply, energy, or space. It requires a discussion of patterns of creation and consumption. The modern conception of the circular economy, on the contrary, is based on overcoming saturated markets and abundance, not a deficiency. The circular economy is ecological because its activities are regional, low carbon, and they use few resources, preserving water, energy, and materials. In this regard, designing buildings with minimum resource consumption during use is becoming mandatory in several European countries. This method represents new challenges to architects, designers, and engineers and promotes the "throwaway linear culture" into a circular system likened to the ecosystems located in the natural environment (Fig. 4.1) (Baker-Brown, 2017; Till, 2012).

In the process of the circular economy, three approaches, useful to reinsert the ignored materials in circulation, assume notable importance: "recycling," "reusing," and "using less." These are the most common strategies that take part in the waste hierarchy, a tool used in the evaluation of methods that protect the environment, adjacent resource, and energy consumption from most beneficial to least beneficial actions. The hierarchy establishes favored program priorities based on sustainability, emphasizing the importance of waste minimization (European Commission, 2008; Hansen et al., 2002) (Fig. 4.1). It is useful to define the most common approaches of the waste hierarchy to understand the subtle differences between the strategies to adopt:

I. **Recycle**: It is any recovery process, which reprocesses waste into materials or products, for the original or other purposes. It includes composting but not incineration. Although the process reduces the quantity of waste, considering it as a valuable stock, it involves the consumption of a consistent amount of energy and water.
II. **Reuse:** It means giving the products a second life before they become waste. There is not a reprocess of the discarded material and, consequently, it does not imply a significant carbon footprint associated with the recycling operation. The value of the waste resource can stay the same or it can increase. This method requires creative designers who recognize the potential in the discarded materials. In this case, one also refers to "*superuse*," a method that differs from the simple "reuse" for the need to propose future scenarios for the rejected material.

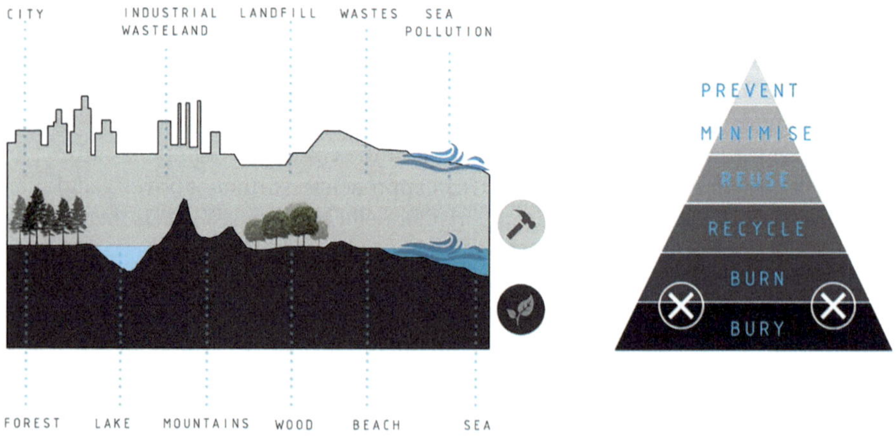

Fig. 4.1 Mining the Anthropocene and waste hierarchy. (Katia Talento, 2020. Partially elaborated from Baker-Brown, 2017)

In this way, it becomes more beneficial than before considered (Use It More. Superuse Studios, n.d.). The practice of reuse comprises three levels of preparation for secondary use:

- *Direct reuse:* The elements are used as close as possible to their original status and for their original purpose, needing almost no preparation.
- *Renewed use*: The operations of cleaning, renovating, repairing, or remanufacturing lightly alter the materials to serve a new function.
- *Rethought reuse*: Reclaimed materials are combined with others to produce a secondary good with a new utility (**Temporal Process in Research, Green Building and Material Reuse.,** 2011).

The advantages of reusing components can be found in various sectors:

- Design: The creation of new materialism.
- Environment: Reducing waste disposal, gas emissions, and climate change.
- Economy: The local stock of resources leads to an improvement in local employment.
- Resource conservation: Possibility of creating materials from primary resources.
- Social: The birth of a new approach of respect toward the built surroundings.

III. **Using less**: Ability to minimize the amount of material used during their whole life cycle.

In this regard, it becomes essential to mention the opinion of the architect Jeremy Till:

> As we move from the industrial age to the digital age, this change provides an opportunity for a new way of thinking about materials. Products should not have a life but should be part of an ongoing technical cycle. (Till & Schneider, 2012).

Being the main topic of the research and the operation being more advantageous from the economic and energetic point of view, the strategy of reuse represents the core of the current chapter.

4.3 Unwanted Material and the Practice of Reuse

The idea of using the place and the surrounding area as the first spot to look for resources is extraneous to most contemporary designers. However, in the past (and still today in specific parts of the planet), discarding materials was not an option, since working with new matter appeared expensive (Gorgolewski, 2018). Although episodic, the practice of reuse has always accompanied the history of building. Since the time of our hunter-gatherer ancestors 10,000 years ago, humanity has always had problems finding resources. People have not stopped using and wasting resources, and living sustainably does not merely impact human societies. For this reason, contemporary society is in the middle of the mass extinction of species, and several resources are limited and hard to access (Baker-Brown, 2017).

Kevin Lynch, in his treaty on waste, provided us with some examples of waste reuse since antiquity. He has declared that in ancient cities, old monuments were used as quarries for stones, beams, or roofing materials. A Roman Imperial script, in 397 BC, authorized the oriental Prefect to use the material of the demolished Pagan temples for the maintenance of public bridges, streets, aqueducts, and shafts. An additional case is related to the Great Fire of London where an enormous amount of old rubble, transported by ship as filling material, created the new Russian city of St. Petersburg (Lynch, 1992). For the construction of Notre Dame du Haut chapel in Ronchamp in the 1950s, even Le Corbusier decided to include masonry walls built from stones recuperated from another ancient chapel (Gorgolewski, 2018).

Since the Industrial Revolution, the consumption of natural resources has risen significantly and unsustainably. Nowadays, nearly 90% of raw materials are destined to become waste, while 80% of products are dumped within the first 6 months of their life (Girling, 2005). This conception is crucial to realize that it is necessary to reuse the discarded material to limit its capacity on the territory (Lynch, 1990). The construction industry, succeeded by the mining and quarrying industries, is still the most abundant source of waste annually generated in Europe, producing 36.4% of discarded material in the first case, and 25.3% in the second one (Eurostat Statistics Explained. Waste Statistics, 2019).

Generally, all human consumption, in every economic activity, is annually responsible for almost 2 billion tons of waste, which increases year after year as populations and urbanization grow in all corners of the world (Waste Atlas, n.d.). The actual world population (7 billion) is expected to rise by 41% by 2050. The problem is that the world does not have enough natural resources, and there is a need to alter human demand, instead of discarding things (D. Cohen, 2007). In a society confronted with climate change, enhanced resource scarcity, and other environmental, economic, and social difficulties, access to new resources and the disposal of scraps are becoming far more expensive and restrained. For this reason, it is essential to "absorb" the enormous piles of wastes, accumulating on the territory, and the rising cost of disposal. From these data, it is feasible to create new and different business opportunities without damaging natural resources and thinking about architecture as a tool to "discover" local and pre-used materials, not always associated with new technologies and the most innovative materials (D. Cohen, 2007).

The modern mind supports the idea based on the change of the world. Modernity is to refuse its actual state and to decide on transforming it to reduce waste. By the turn of the twentieth century, waste represented the most significant problem of society and, during the 1980s and 1990s, recycling and reusing programs began to rise in the industrialized world. At this time, the waste products exposed a different story, that of wasted resources (Hebel et al., 2014). Some European countries, such as Belgium, the Netherlands, and Denmark, recycle/reuse between 95% and 97% of their waste. Denmark represents the most significant case thanks to the approval of a law in 1988 that imposed to the building companies the reuse of the discarded material from excavations, demolitions, and constructions. This country has been a front-runner in investigating the implications of circular practices, recognizing that the products/buildings in a system need less energy, to produce fewer gas emissions and diminish the demand for raw materials.

This premise is essential to understand that the contemporary world is beginning to consider unwanted material not as a separate entity but as a component connected with the building sector. According to this logic, waste is the consequence of any human action and interaction, bringing natural raw material from one stage of being into another, by applying different forms of skills and energy. As the architect Joachim Mitchell affirms, the discarded matter is understood as an integral part of a resource, and the future city makes no distinction between waste and supply (Mitchell, 2013).

The waste might become so attractive that it makes reason to grow more of it (Hebel et al., 2014). For thousands of years, cities were designed to produce waste; nowadays, on the contrary, it is the time to design waste for the regeneration of cities.

How could waste and urban development combine?

The first approach to consider is the reduction, thinking about a radical reuse plan (Rogers, 2005).

According to the first step of the research methodology followed in this chapter, meticulous and selective bibliographic research has elaborated the waste reuse theoretical notions (described so far). The references have been searched for keywords, mostly related to the concepts of waste, waste reuse, and stone wastes. As happened for the previous chapters, the bibliography was then organized following the research purpose, namely that of generating a fluent reading about the waste concept, from its origins to its contemporary applications. As described in the following sections, this solid background represents the starting point to discuss the waste reuse strategy in the practical sector, which can occur using two distinct tactics: off-site reuse and in situ reuse.

4.4 Building with Stone Wastes

4.4.1 Off-Site Reuse Applications in Architectural, Landscape and Art Sector

Nowadays, different types of wastes exist around the world. Among the most popular, one can find plastic, paper, glass, radioactive, medical, biological, chemical, industrial, solid, hazardous, agricultural, and demolition and construction wastes. They represent distinct rejected materials with specific properties and origins. The common factor that connects all these wastes is the possibility to re-introduce them in a new life cycle, generating at the same time new functions and uses, instead of demolishing them. It is also evident that not all waste materials are suitable for this approach. Chemical, radioactive, and medical waste products need special treatment, and they are not a part of the current research. On the contrary, the research topic pays attention to industrial wastes, especially those from quarries that have lost their meaning.

In the architecture field, among the innovations of the constructive culture regarding the alternative use of wasted stone, it is necessary to mention the "gabions technology," which transfers a technique, originated, and well established in the world of water and road works, to the architectural and landscape field. This system is primarily recurrent in landscape design and numerous projects of quarry reconversion. The most ancient testimonies relate to the use of mats and vegetal baskets, for the transport and consolidation of piles of stones and ground, dating back to Sumerian, Babylonian, and Egyptian civilizations. Since the Modern Age, this method spread to the construction of military and waterworks, introducing the use of wire containers. If in the past gabions were produced in situ, nowadays they are prefabricated and appropriate for many architectures and landscape uses (Acocella, 2007).

One of the first applications concerns the project of the architect Enric Miralles of the **Igualada Cemetery** (Fig. 4.2). It represents the regeneration of a former extractive area where the designer introduces the use of gabions technology as self-supporting walls. In this way, one defines both a containment system and facade configurations, "imprisoning" salvaged stones of the quarry, using a wire netting. Other recovered elements, evocative of waste processes, are embedded into paving. Beyond the powerful landscape integration, gabions have notable quality performance related to the operation of reuse and recycle.

Fig. 4.2 Igualada Cemetery, Spain. (Wikipedia, 2020)

In this regard, the architect Alfonso Acocella affirms that gabions are permeable to air, and, at the same time, they have high thermal inertia. They are inexpensive, easy to transport, and durable. Moreover, gabions do not require maintenance; they are modular, removable, and can be reused. Furthermore, they can also contain recycled material as filling, since their first fabrication (Acocella, 2007). Many other projects, from different architects and with different purposes, apply the technology of gabions to limit the enormous amount of waste located on the territory from extraction activities (Figs. 4.3, 4.4, 4.5 and 4.6).

This technology denotes the most popular and practiced type of reuse of stone waste from quarries. It mostly represents "off-site" reuse, which means to take the discarded material from the extraction sites and to transport it outside the area where it resides, to create new components for architectural building or landscape elements. Beyond the impressive gabions approach, stone wastes are reused for the composition of building facades and walls or the covering of internal spaces. In the first case, there are many ways to design building facades with residual rocks, and it can happen using only stone without containers and binders or panels composed of the residual substance.

Mortensrud Church represents an exemplar case regarding the realization of building facades, composed of residual stones without the use of wire netting or mortars (Fig. 4.7). The church stands on the top of a hill, in a dense pine forest outside Oslo, and it was designed by Norway's Jensen and Skodvin Architekten.

Fig. 4.3 Adega Dominus, Napa Valley. (Wikipedia, 2020)

Fig. 4.4 Long Peak Toilet, Colorado. © Colorado Building Workshop/Rick Sommerfeld

Fig. 4.5 Private House, Stadtbergen (on the right) and social housing in Montpellier (on the left). (Wikipedia, 2020)

Instead of removing the sharp rocks and the irregularities existing in the place, the architect's concept centered on including these natural elements into the building. The main structure of the building is a framework of fixed steel profiles, stabilized by horizontal beams. The walls are a mixture of slate stones and glass, with the slate

Fig. 4.6 Discovery of three Roman industrial furnaces, Spain. © Adrià Goula

Fig. 4.7 Mortensrud Church, Norway (on the right). (Wikipedia, 2020). Apartment 1, Iran (on the left). Architecture by Collective Terrain Architect: Ramin Mehdizadeh. ©Omid Khodapanabi

walls withdrawn from the glass facades, allowing for narrow passages on either side of the nave. The stone in this wall is built without mortar, thus letting in light, and it has one regular side and one irregular as standard. The stone wall is horizontally set by steel plates that span between the columns, inserted into the wall, every meter (Mortensrud Church / Jensen & Skodvin Architects, 2011).

It is feasible to create waste building facades also through the formation of covering panels with stone scraps. A project which deals with this approach is **Apartment 1** by the architect Mehdizadeh in Iran. It represents an apartment building, with commercial spaces, entirely produced with waste from local quarries to respect sustainability principles in the architecture field (Fig. 4.7). Due to the intense extraction activity, the designer, having available an enormous quantity of scraps, decided to stimulate local players toward energy savings and reduction of global costs of construction. He designed exterior covering panels by overlapping different types of waste, such as marble, travertine, and granite. In this way, the architect created high surfaces with mixed chromatic and tactile nuances (Frearson, 2013).

The concept of covering building surfaces with stone waste can also apply to the interior spaces of a construction. **Valentino New York flagship store**, designed by the architect David Chipperfield, expresses one of the most outstanding projects where the abovementioned suggestion is practiced. By merging old and new, the store idea makes a palazzo-like atmosphere, steering away from a showroom appearance of a typical boutique and developing a more architectural retail format instead. Architectural components are at the heart of the design notion. Architecture is brought into the shop rather than persisting only on the exterior, thereby decreasing the use of superfluous decoration and thin panels to generate an interior scenography with the use of recycled marble waste for the covering of walls (Fig. 4.8). Architectural characteristics, such as columns, staircases, and walls, continually expose their fundamental materiality, and they complement the pieces on display (Howarth, 2014).

The technology of reusing stone wastes to limit their impact in the territory is also present in the landscape field for different purposes. Wastes are the best means to enhance the meaning and the memory of a place. Raw stones appear to be the best example of materials to reveal the history of a place, jointly with its mutations and "sorrows." Memorial parks often manage this type of scrap to emphasize their past. A relevant case is the **Belzec Memorial Site** in Poland. This project was founded in 2004 with the responsibility of documenting and commemorating the Jewish victims of the German death camp in Bełżec. The memorial site, prepared by a group of artists, exposes the cold and brutal strength of the place on the border of life and death. It is a site where all the arts converge, where the scale is gigantic, but at the same time, the artistic language is reserved, free of naive statuary or Motherland symbols. A large rectangle of several hectares represents the cemetery without tombs, and it is entirely covered by a layer of specially prepared material (stone wastes mixed with cinders and sterile soil), which creates a black spot in the green Roztocze Lubelskie countryside (Fig. 4.9) (Knight, n.d.).

Through this strategy, the landscape sector takes advantage of waste also to magnify the industrial memory of a place. It is the case of **Piedra Tosca Park**, located in the Natural Park of the Garroxta Volcanic Zone. The basaltic flow from the volcano partially generated the area. The strenuous work of humans to cultivate a small portion of land, accumulating all the rocks, stones, and gravel in deep walls, tombs,

Fig. 4.8 Valentino New York flagship store. ©Santi Caleca

and barracks, also contributed to the formation of the park. It is a particularly unique site, a sea of wasted rocks (Fig. 4.10) ('Piedra Tosca Park/ RCR Arquitectes', 2016). Other applications are detectable in the landscape field, such as those related to public parks and outdoor items of furniture. For these purposes, one usually uses

Fig. 4.9 Belzec Memorial Site, Poland. ©Giovanni Carrieri

Fig. 4.10 Piedra Tosca Park, Natural Park of the Garroxta. ©RCR Arquitectes

gabion technology (as previously mentioned) to create gardening elements such as paths, partition walls, seats, flowerbeds, and fountains.

The art sector was the pioneer of the concept of reusing waste, thanks to the work of land artists who have recognized the importance of the shared topic. In 1967, Robert Smithson explored the industrial areas around New Jersey, and he painted them as the equivalent of antiquity monuments. Captured by the swirling pattern of brittle rocks, the artist found in those wasted areas a petrified sea with many stratigraphic horizons and endless directions of steepness. Following his idea of art based on the notion of limit, Smithson imagined containing that "oceanic site" through a sculpture; installed in 1968, a trapezoidal wood box enclosed the uneven fragments of the slate quarry in Bangor, Pennsylvania (Smithson, 1996). He described the installation as a three-dimensional map of the quarry, a piece of an already fragmented landscape. This work, titled *Non-site*, is one of the numerous *Non-sites*, which the artist created that year. In each of them, he carried matter from extraction sites, especially mines and quarries, and placed them as sculptures, adjacent maps, and other documentation of the original place, into galleries. Despite the location in gallery spaces, the installations represented means for dialectical thinking, and they projected to the place and back again (Smithson, 1968).

These pieces referred to the spot that they came from and to their human, historical, and geological mutations, and they were not only fixed objects. Material relocations and their diversity connected the sites and non-sites. Sites were physical and peripheral and supplied means for urban development; *non-sites*, on the contrary, were abstracts and central concentrations of cultural capital (Alloway, 1983). Sites were the "signified" and *non-sites* were the "signifier." The conception of the land artist Smithson examines, in the art field, how the production of the landscape in one area is related to its alteration elsewhere, by exalting the place of material production and denying the place of consumption (Hutton, 2019). According to these considerations, all the examples described above, consider the wastes as elements to use elsewhere, aiming to exclude them from their original location. In the next chapter, the research directions suggest acting in situ, not exporting the material.

4.4.2 Reuse Applications in Situ

Today, although there is a positive willingness to apply reuse strategies for the waste material (coming from quarries and other extraction sites) to decrease its repercussions on the territory, this is not enough. The amount of waste is so severe that the operations of reuse cannot involve all of it but only small percentages, compared to the totality of wastes. Industrialization has advanced so far that, on the one hand, human activities have taken over the natural aspects of the environment, and, on the other hand, they do not know how to manage the industrial leftovers that cannot be reused.

Based on this evidence, what happens to the remaining part of wastes?

Practical answers to this question do not exist. The unique alternative is to accumulate the leftover material in giant heaps since the hypothesis of demolishing it is unaffordable from an economic point of view. In this way, one treats the mounds of waste as burdens and abandoned components, the function of which is only that of defacing the landscape. They seem anonymous elements, not communicating with the landscape, and refused by society due to their extraneous aspect. Discordant with this attitude, the current research attempts to propose working with the artificial mountains of industrial waste from extraction sites because they are the most intricate and fragile part of the territory. Considering these accumulations as indispensable components of the landscape represents a challenge, and it involves future extensions of the cities, reimaging the discarded elements of our wastelands. The giant piles need to be re-introduced in a cycle, from a landscape and architectural point of view. The inert matter is not an obstacle to overcome, but an occasion to restart and remodel the territory with the contributions of architects and planners.

Moreover, they also could be innovative elements to generate new economies for society. After providing a background on the waste reuse topic, the current investigation intends to put at the center of the architectural discipline the contemporary problem related to the presence of enormous mounds of stone remains in our landscapes. The architecture field has the duty of considering these elements in the process of rehabilitation of the territories, jointly with the quarries' (and other extraction sites') regenerations. The waste heaps cannot go unnoticed because, following human transformations, they turn into contemporary landmarks of the cities. Skyscrapers, towers, cathedrals, memorials, monuments, bridges, installations, and sculptures are not the only ones. Landmarks are also those produced by the industrial activity, composed of waste, which are becoming integral pieces of our environment, being repositories and testimonies of technical, human heritage, and local identity.

For this reason, as will be described in the next chapter, the present investigation contributes to this innovative theme, by analyzing a massive area in Portugal constituted of numerous giant waste heaps from marble quarries (Figs. 4.11a,b, Table 4.2) belonging to the Estremoz Anticline, in the Alentejo province. Before conceiving an architectural approach and a renovated function to the artificial accumulations of wastes, it is essential to realize what exists concerning these "unfamiliar" industrial elements. The bibliographic search is extremely complicated due to the absence of sources and records, and most documented waste piles are mainly originated from the mining activity. Most of them appear to be disregarded and abandoned in the landscape, "silent" and deprived of a role. Despite being massive and impressive, they only continue to represent the background of the urban or rural scene. Mining activities are the primary producer of accumulated wastes, followed by wastes extracted from quarries. As illustrated in the following image, these accumulations, even without a function and recovery interventions, embody the aspect of new contemporary monuments, which deserve protection measures (Figs. 4.11a–i).

Due to the complexity of going beyond the traditional limits in the architectural context, the reuse interventions of a mound of waste material are considerably restricted. Because of the lack of bibliographic sources in the architectural domain,

Fig. 4.11 Waste heaps and elaboration of a general matrix with distinct hypotheses of reuse applicable to the piles of remains. (Katia Talento, 2020)

it is more complicated to trace few examples in which the large piles assume new meaning. Until now, the operations of reuse rely on ephemeral or small interventions, and they mainly include recreational and touristic uses to promote the industrial heritage. The community of Blaenau Ffestiniog, located in the heart of **Snowdonia National Park** in North Wales, represents an exemplar case, which

Table 4.2 Elaboration of new typologies for the reuse of waste heaps and synthesis of the discussed case studies. (Katia Talento, 2020)

Typology	Discussion	Case study	Type of waste	Use
Detachment	It presupposes the insertion of *small ephemeral structures in proximity to waste piles, not interfering with them. There are respect and a valorization of the preexistence by contemplating it*	*Blaenau Ffestiniog, North Wales.*	Slate	Touristic
Equality	*It consists in treating the accumulation of scraps as the other components of the landscape. The waste accumulation represents not an isolated factor to monumentalized but a territorial component able to interact with people*	*Weldon Spring Centre, St. Charles County, Missouri*	Nuclear	Recreational, touristic
Binomial monumentality-human interaction	*It includes both monumentality and interaction components, exalting the waste accumulation as a prominent installation, opening it to people's visits*	*Dionyssos quarries, Greece*	Marble	Touristic
Naturalistic	Mimesis with the landscape. *The waste heap is treated as an integral part of the landscape and not as an extraneous object to contemplate. This criterion allows the introduction of recreational and cultural structures only if they appear "hidden" in the waste accumulation*	*Slag heap Haniel, Germany*	Coal	Recreational, touristic
Architectonic	It completely exalts the meaning of the pile of remains giving it the function of "shell." the waste pile can contain buildings and renewed space	*Temppeliaukio church, Helsinki*	Mix of solid blocks and granite wastes	Religious

pays attention to the industrial memory of the place, valorizing it as much as possible. The park, besides being recognized for its lofty peaks, is also known for its magnificent scenery, composed of vast slate mines, caverns, and massive waste heaps (Gallent et al., 2017). In this locality, an attractive touristic system has been designed, and safari tents, immersed in the industrial landscape, which observe the industrial remains, form it (Figs. 4.11l,m). This type of intervention consists of introducing small ephemeral structures in proximity to mines and slate waste piles, not interfering with them. If, on the one hand, it can be interpreted as a timid approach, on the other hand, it respects and valorizes the preexistence by contemplating it.

A different strategy of reuse is that of physically working on the waste heap through little interventions. **Weldon Spring Centre**, located in St. Charles County, Missouri, once was one of the most toxic places in the United States. During the Second World War, the American Government built a munitions plant on the property. Old bunkers used for the storage of TNT (trinitrotoluene) explosives still dot the landscape. Later, US Environmental Protection Agency (EPA) Superfund used the site to reprocess TNT and uranium and decontaminate the land (Office of Legacy Management, n.d.). Despite that, the giant mound of nuclear waste persists in the territory, and it was equipped with staircases to permit visitors climbing it and capturing beautiful views of the surrounding area (Figs. 4.11n,o). Nowadays, the conservation area includes forest, woodland, cropland, grassland, sand flats, lakes, ponds, and wetlands. It has developed into a valuable wildlife habitat, and it offers a wide variety of recreational opportunities for tourists, including fishing, hunting, hiking, and biking. Albeit in an embryonal form, this method shows the desire to treat the accumulation of scrap like the other components of the landscape. According to this view, the waste accumulation represents not an isolated factor to monumentalize but a territorial component able to interact with people.

The union of these two elements, monumentality and human interaction, is interlaced in the **Dionyssos Quarry** in Greece, always used for the extraction of marble. Following the idea of the creators, that of "celebrating" marble quarries, they conceived an open-air museum to safely observe the extraction activity. According to a fascinating artistic vision, the landscape sculptor Nella Golanda and the architect Aspassia Kouzoupi treated the quarry as an impressive sculpture thanks to the "absorption" of the waste embedded in the pit. As a result, the waste faces had become a sculptural composition that revealed the lines of the rocks. The museum consists of a series of small structures and paths through the old quarry, built without the use of mortar or concrete. The designers also wanted the site to contain surprise factors such as the climb ends in a hollowed belvedere with a marvelous view of the sea (Figs. 4.11p,q). Applying conventional industrial methods, they have produced a post-industrial landscape of sublime richness (Holden, 2003). If the previous examples were separately considered, in this case, Dionyssos Quarry includes both monumentality and the interaction components, exalting the place as a prominent rock installation, opening it to tourist excursions.

Slag Heap Haniel, in Germany, adopted a more naturalistic approach. It is the highest elevation in the Ruhr region (Chmielewska & Otto, 2014), and it appears camouflaged in the surrounding landscape, resembling a natural hill of the territory. This heap is treated as an integral part of the landscape and not as an extraneous object to contemplate. For this effect, the reuse process consists in the creation of a large park, in the southern part of the heap, while the northern one continues to be used by Prosper Haniel Coal Mine. The waste pile is blended in the context, and the recreational structure for cultural events and the amphitheater are placed in the "crater" of the slag heap (Fig. 4.11r). It is hiding on the top of the "mountain," inside an artificial "valley." In this manner, Slag Heap Haniel is suitable for leisure and entrainment functions with respect of the naturalistic aspects.

Beyond the recreational and touristic purpose, there was an attempt to reuse part of the waste, accumulated in piles, also for different and more specific functions, such as religious functions. In Helsinki, an unusual building, **Temppeliaukio Church** (also known as Stone Church) was built from a combination of solid rock and waste blocks carved out of a granite stone hillside (Fig. 4.11s). The domed construction is not inserted within an industrial context but at the center of the city where there are no quarries, mines, and enormous mounds of waste in the location. The granite wastes are united with the solid block of stone for the creation of a new structure inside it. Due to the peculiarity of the project, the research also considers this reference. Building inside a combination of rocks and wastes could represent a basis for future investigations to apply to the mounds of waste from quarries.

This schematization represents the application of the theory about waste reuse and circular economy, previously described, in the practical field, through the selection of significant case studies in which the attempt to reuse the waste heaps (albeit in an embryonal form) is evident. The matrix also embodies the figurative synthesis of the research methodology followed in the work, which contains both the theoretical and practical foundations for deriving, in the next chapter, the systematic architectural methodology for the rehabilitation of quarries and the reuse of waste heaps. Being a theme with a lack of fundamental and applied research and having available a limited quantity of sources, it was difficult to search for case studies that show the process of reuse and conversion of the waste piles. Consequently, it was easier to select a set of cases among those found.

As for the theoretical section, the best examples of reuse have been identified by keywords, mostly related to the concepts of waste reuse, mine/stone waste, and quarry regeneration. Successively, the designated case studies have been differentiated in typologies of intervention (off-site and in situ), deepening that in situ, being part of the main purpose of the research. From this selection, the existing approaches of reuse have been investigated to compose a general matrix with distinct hypotheses applicable to the piles of remains. The obtained results are divergent from those already published. While the majority of the literature is focused on the recycling of the waste material in the industrial sector (Deepankar, 2018; Silva & Esteves, 2015; Simão et al., 2016; Ulubeyli & Artir, 2015), the current scientific method shows the possibility of approaching the practical reuse of waste heaps, in the architectural and landscape sectors, through the definition of typologies of reuse (Table 4.2).

From the elaboration of typologies of reuse, it is evident that the selected waste piles have been mainly reused in situ for recreational and touristic purposes. Most of them have been considered monumental containers of spaces (both indoors and outdoors), integrated into the landscape, and always communicating with the preexistence and its meaning. These main concepts have been crucial to deductively derive a systematic architectural methodology, based on the local reuse of the abandoned industrial waste heaps and the regeneration of quarries.

As will be discussed in the following chapter, the proposed new methodology represents the main outcome of this research, applied to a specific location. Therefore, this work adds an important new contribution to the field and an advance to the state of the art. The elaborated systematic methodology identifies and organizes the main examples of waste piles, to formulate new hypothesis and general criteria of reconversion, resolving the existing gap in both architectural and theoretical domains. To the best of the author's knowledge, there is not any similar schematization in the literature, and the idea of dealing with the mounds of waste in the architectural and landscape sectors has been slightly approached until now.

4.5 Chapter Resume

This chapter addressed the problem associated with the massive accumulations of waste from extraction activity. These products are usually excluded from the process of rehabilitation because they are considered disturbing elements without economic and architectural value. Recently, greater awareness toward environmental concern has considered the hypothesis of introducing waste material in the procedure of the rehabilitation of quarries. In many places, intervention actions are being developed, but they are carried out in a disjointed manner and without a dedicated support methodology that defines and systematizes the criteria to adopt. Through architectural interventions, the rehabilitation of quarries and landscapes can obtain a more significant advantage, transforming the waste heaps into new spaces with renewed functions.

According to the research, this chapter has set up the basis for the elaboration of a practical methodology of reuse to apply to a case study in Portugal. As described in the next chapter, this methodology represents a starting point to investigate innovative solutions for the rehabilitation of quarries and the reuse of waste heaps from the extraction industry. The inclusion of the inert material in quarry rehabilitation represents an advantage that can be addressed by the architecture discipline to transform and innovate the landscape, respecting the principles of the circular economy. Giving a new life cycle to the abandoned components of the territory has repercussions on the local economy and the management of the land, which highly contributes to a sustainable use of the planet and to further improve the concept of sustainability and circular economy.

Bibliography

Acocella, A. (2007). *Architettura dei gabbioni*. Costruire.

Alloway, L. (1983). Sites/Nonsites. In *Robert Smithson: Sculpture* (pp. 42–43). Smithmark.

Angélil, M., & Siress, C. (2010). Re; going around in circles. In *Re-inventing construction* (pp. 248–264). Ruby Press.

Baillie, P. W. (1992). Auditing a flooded quarry prior to marina development. *Environmental Management, 16*, 531–540. https://doi.org/10.1007/BF02394129
Baker-Brown, D. (2017). *The re-use Atlas. A designer's guide towards a circular economy*. RIBA Publishing.
Bauman, Z. (2004). *Wasted lives: Modernity and its outcasts*. Polity Press.
Bryony, R. (2016). *Tabula Plena. Forms of urban preservation*. Lars Muller Publishers.
Cheyne, I., & Purdue, M. (1995). Fitting definition to purpose: The search for a satisfactory definition of waste. *Journal of Environmental Law, 7*(2), 149–168.
Chmielewska, M., & Otto, M. (2014). Revitalisation of spoil tips and socio-economic polarisation–A case study of Ruhr area (Germany). *Environmental & SOCIO-economic Studies, 2*(2), 45–56.
Chung, S. S., & Poon, C. S. (2000). A comparison of waste reduction practices and the new environmental paradigm in four southern Chinese areas. *Environmental Management, 26*, 195–206.
Cohen, D. (2007, May). *Earth's natural wealth: An audit*. https://www.newscientist.com/article/mg19426051-200-earths-natural-wealth-an-audit/?ignored=irrelevant
Dal Sasso, P., Ottolino, M. A., & Caliandro, L. P. (2012). Identification of quarries rehabilitation scenarios: A case study within the metropolitan area of Bari (Italy). *Environmental Management, 49*, 1174–1191. https://doi.org/10.1007/s00267-012-9847-0
De Poli, M., & Incerti, G. (2014). *Atlante dei paesaggi riciclati*. Skira editore.
Deepankar, K. A. (2018). Feasibility of waste marble powder in concrete as partial substitution of cement and sand amalgam for sustainable growth. *Journal of Building Engineering, 15*, 236–242. https://doi.org/10.1016/j.jobe.2017.11.024
Douglas, M. (1966). *Purity and danger*. Routledge and Kegan Paul.
El-Fadel, M., Sadek, S., & Chahine, W. (2001). Environmental management of quarries as waste disposal facilities. *Environmental Management, 27*, 515–531. https://doi.org/10.1007/s002670010167
European Commission. (2008). Directive 2008/98/EC on waste (Waste Framework Directive). https://ec.europa.eu/environment/waste/framework/
Eurostat Statistics explained. Waste Statistics. (2019, June). https://ec.europa.eu/eurostat/statistics-explained/index.php/Waste_statistics
Frearson, A. (2013, May 10). Apartment no. 1 by AbCT. *Dezeen*. https://www.dezeen.com/2013/05/10/apartment-no-1-by-abct/
Freilla, O. (2008). *The south Bronx: A neighbourhood reclaimed in actions: What you can do with the city*. Canadian Centre for Architecture.
Gallent, N., Mace, A., & Tewdwr-Jones, M. (2017). Mainland Europe and Britain compared introduction. In *Second homes* (pp. 127–143). Routledge.
Girling, R. (2005). *Rubbish! Dirt on our hands eand crisis ahead*. Eden Project Books.
Gorgolewski, M. (2018). *Resource salvation: The architecture of reuse*. John Wiley & Sons Ltd..
Gourlay, K. A. (1992). *World of waste: Dilemmas of industrial development*. Zed Books. https://doi.org/10.1177/027046769301300444
Hansen, W., Christopher, M., & Verbuecheln, M. (2002, December). *EU waste policy and challenges for regional and local authorities*.
Hebel, D. E., Wisniewska, M. H., & Heisel, F. (2014). *Building from waste: Recovered materials in architecture and construction*. Beirkhäuser. https://www.researchgate.net/publication/295519671_Building_from_Waste_Recovered_Materials_in_Architecture_and_Construction
Holden, R. (2003). *New landscape design*. Architectural press.
House of Commons Environmental Audit Committee. (2014, July). Growing a circular economy: Ending the throwaway society. *Third Report of Session* 2014–2015.
Howarth, D. (2014, September 11). David Chipperfield's Valentino flagship store opens in New York. *Dezeen*. https://www.dezeen.com/2014/09/11/david-chipperfield-valentino-flagship-store-fifth-avenue-new-york/
Hutton, J. (2019). *Reciprocal landscapes: Stories of material movements*. Routledge.

Knight, E. (n.d.). A reflection on Belzec. The National Holocaust Centre and Museum. Retrieved 17 March 2020, from https://www.holocaust.org.uk/blog/belzec
Lynch, K. (1990). *Wasting Away*. Sierra Club Books.
Lynch, K. (1992). *Deperire. Rifiuti e spreco nella vita di uomini e città*. CUEN.
Mitchell, J. (2013). Turning waste into building blocks of the future city. *BBC Online News.*. https://www.bbc.com/future/article/20130524-creating-our-cities-from-waste
Mortensrud church/Jensen & Skodvin Architects. (2011, August 31). Archdaily. https://www.archdaily.com/1929/mortensrud-church-jsa.
OECD. (1998). Final guidance document for distinguishing waste from nonwaste.
Office of Legacy Management. (n.d.). About Weldon Spring site interpretive center. Energy.Gov. https://www.energy.gov/lm/about-weldon-spring-site-interpretive-center
Oxford English dictionary online. (n.d.). *Waste*. https://krygier.owu.edu/temp/waste_oed.pdf
Piedra Tosca Park/RCR Arquitectes. (2016, April 7). Arch Daily. https://www.archdaily.com/806228/piedra-tosca-park-rcr-arquitectes.
Pongrácz, E., & Pohjola, V. J. (2004). Re-defining waste, the concept of ownership and the role of waste management. *Resources, Conservation and Recycling, 40*(2), 141–153. https://doi.org/10.1016/S0921-3449(03)00057-0
Rhys, J. (2011). *Cumulative carbon emissions and climate change*. Institute for Energy Studies.
Rogers, H. (2005). *Gone tomorrow: The hidden life of garbage*. The New Press.
Silva, C. F., & Esteves, L. D. (2015). The mounds of Estremoz marble waste: Between refuse and reuse. In *Waste 2015–Solutions, treatments and opportunities* (pp. 305–310). CRC Press.
Simão, J., Freire, A. C., & Farto, H. (2016). *Caracterização de rejeitados de escombreiras de pedreiras de rocha ornamental para aplicação em camadas não ligadas de pavimentos rodoviários*. 1–9.
Smithson, R. (1968). A provisional theory of non-sites. In *Robert Smithson: The collected writings* (p. 364). University of California Press.
Smithson, R. (1996). A sedimentation the mind: Earth projects. In *Robert Smithson: The collected writings* (pp. 110–111). University of California Press.
Steffen, A. (n.d.). *Quotefancy*. https://quotefancy.com/quote/1427273/Alex-Steffen-There-is-no-such-thing-as-garbage-just-useful-stuff-in-the-wrong-place
Talento, K., Amado, M., & Kullberg, J. C. (2019). Landscape—A review with a European perspective. *Land, 8*(85). https://doi.org/10.3390/land8060085
Talento, K., Amado, M., & Kullberg, J. C. (2020). Quarries: From abandoned to renewed places. *Land, 9*(136). https://doi.org/10.3390/land9050136
Temporal process in research, green building and material reuse. (2011).
Till, J. (2012). *How will architects be educated in 20 years time?* www.jeremytill.net/read/97/how-will-architects-be-educated-in-20-years-time
Till, J., & Schneider, T. (2012). Invisible agency. *Architectural Design, 82*(4), 38–43.
Ulubeyli, G. C., & Artir, R. (2015). Properties of hardened concrete produced by waste marble. *Procedia Social and Behavioral Sciences, 195*, 2181–2190.
UNEP. (1989). Basel convention on the control of transboundary movements of hazardous wastes and their disposal. *Journal of Environmental Law, 1*(2), 255–277. https://doi.org/10.1093/jel/1.2.255
Use it more. Superuse Studios. (n.d.). Arch20. https://www.arch2o.com/use-superuse-studios/
Waste Atlas. (n.d.). http://www.atlas.d-waste.com/
Wilson, D. C. (2015). United Nations environment Programme, & international solid waste association. *Global Waste Management Outlook.*
Wong, M. H. (1986). Reclamation of wastes contaminated by copper, lead, and zinc. *Environmental Management, 10*, 707–713. https://doi.org/10.1007/BF01867725
World Health Organization Expert Committee. (1965). *Environmental health aspects of metropolitan planning and development.*
Zero Waste America. (n.d.). *What is Waste?* http://www.zerowasteamerica.org/whatiswaste.htm

Part III

Chapter 5
Case Study Application

The stone that the builders rejected has become the cornerstone.

Mc 12,1-12

5.1 Introduction

Description of the site and its geological aspects, analysis of its original landscape, effects of the industrialization, replanning of the area and proposal of an architecture strategy of rehabilitation.

The largest marble explorations of Portugal are located in the Estremoz Anticline, roughly 50 km northeast of the city of Évora, in the central interior of the country, and 170 km from the city of Lisbon. This geological structure, which has 42 km length by 8.5 km width, extends in a northwest-southeast direction, intersecting the municipalities of Sousel, Estremoz, Borba, Vila Viçosa, and Alandroal. The quarries only exist in three municipalities of the Anticline: Estremoz, Borba, and especially, Vila Viçosa, where the highest concentration of material resides; this is the reason why this area is known as the "marble triangle" (Fig. 5.1).

The mentioned municipalities belong to the province of Alentejo, which is well known to be a region with extensive and hot summers, where the control and the management of the water have normally been crucial in the habits of their populations. Although the use of the water has been continuously rationed over the years, the area close to the Estremoz Anticline has always had a great abundance of it because it is located inside a large aquifer named "Estremoz-Cano System." That characteristic favors a broad land use and the formation of farms, orchards, and gardens, giving rise to a much greener landscape (Fig. 5.2), contrasting with its surroundings, which is dominated by forestry and rainfed agriculture (Baganha, 2015).

In opposition to this natural aspect of the Estremoz Anticline, there is also an artificial side, made by man, characterized by the presence of a huge quantity of marble that represents the symbol of the strong industrialization of the region. The well-known extraction activity of the Alentejo province has started in the Antiquity. It has not been developed regularly, over the years, but it has been intensified since

K. Talento, *Sustainable Buildings from Marble Waste Heaps*,
https://doi.org/10.1007/978-3-031-72146-5_5

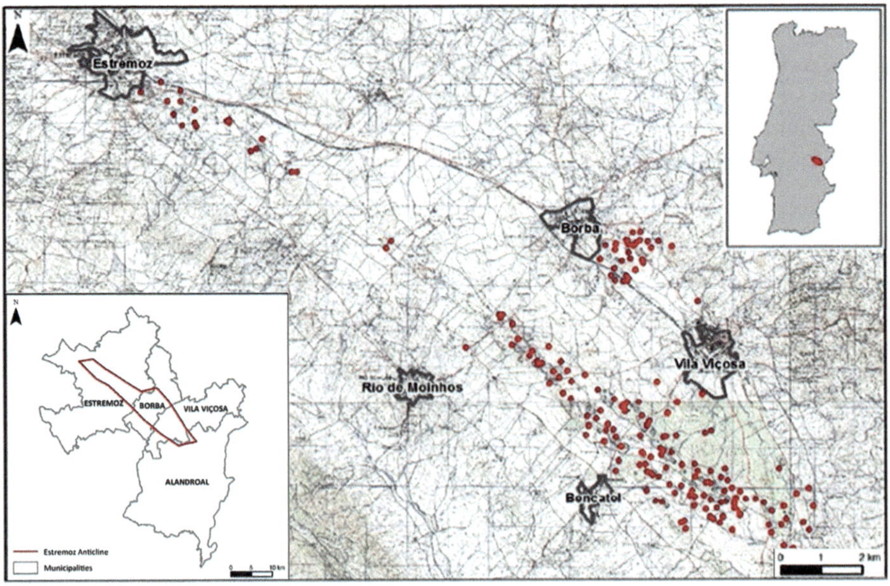

Fig. 5.1 Territorial classification and quarry identification on the territorial map. (Source: PROZOM and Portugal military map, sheets 425–427, 439–441)

Fig. 5.2 Natural landscape of the Estremoz Anticline. (Katia Talento, 2018)

the second half of the last century, with the introduction of new technologies that allowed to improve the efficiency, the speed of extraction and the amount of production of the industrial activity. There was a gradual substitution of the natural energies (such as water, wind, and muscular strength) by new steam, diesel, and electrical ones. In this way, a real modern industry had been created, mainly concentrated in the municipality of Vila Viçosa, which, in a long first period, has also contributed to the exportation of the material abroad (Alves, 2015).

The continuous opening of many quarries gave rise to a landscape of high industrial concentration that has partially overlapped the agricultural one, composed of vineyards, olive groves, cork trees, and cereals. Despite the initial improvements of the economy of the country, the presence of countless quarries has contributed to the degradation of the area and the destruction of the original landscape. After its exploitation, the territory has turned completely degraded due to the presence of several open pits and a considerable number of wastes (Figs. 5.3, 5.4 and 5.5).

After stopping the extraction work, the disabled quarries were neither properly closed, as claimed by the legislation, nor added any value from the patrimonial point of view. Moreover, some quarries were disabled before any legislation was produced, and nowadays, they deal with property issues with the Portuguese regulator.

Due to quality "defects" of the material, of all the extracted marble, only a small percentage (about 9%) is commercialized and the remaining part (about 91%) is accumulated in huge piles of residual stone, named in Portuguese "*escombreiras*" (Figs. 5.6, 5.7 and 5.8).

Fig. 5.3 System of flooded quarries in the Estremoz Anticline (Katia Talento, 2021)

Fig. 5.4 Deep quarry in the Estremoz Anticline. (Katia Talento, 2018)

Fig. 5.5 Deep quarry in the Estremoz Anticline with wastes in the background. (Katia Talento, 2018)

Fig. 5.6 *Escombreira* in the Estremoz Anticline. (Katia Talento, 2021)

The reasons for the accumulation of the residual material are related to the following aspects:

I. The intense fracturing of the rocks, which causes the presence of small and medium blocks, unsuitable for sale
II. The karstification of the marble, which affects the quality and the management of the exploration, largely increasing the quantity of waste material
III. The existence of spots, which drastically reduce the commercial value of the stone, making it unusable

For all these reasons, the marble is only used in a small amount in several industries, and its applications span from the glass, chemical, pharmaceutical, environmental, agricultural, plastic, steel, metal, textile, and construction industry to public works (Ladeira, 1977; Manuppella et al., 1981; Shirazi, 2011). The few applications that can lead to a larger consumption of the waste material are those oriented to civil construction industries, to produce lime and cement, for calcining acid soils in the agriculture sector, and for the composition of bituminous mixtures, useful to road constructions and aerodromes in the civil engineering sector (Costa, 1992; Ribeiro et al., 2010; Ulubeyli & Artir, 2015). The remaining part of the residual material persists on the territory to "deface" the landscape, since the big mounds of wastes

Fig. 5.7 *Escombreira* in the Estremoz Anticline. (Katia Talento, 2018)

cannot be disposed elsewhere due to economic limits. The costs that can be added either in relation to the transport of this residual material to landfills, nearby or far away, or in relation to the work of the individual manufacturing operators are considerable.

In the following sections, the research will focus on the architectural field and, at the same time, having an interdisciplinary approach, it will also explore scientific domains related with landscape and geology. In addition, the architectural and landscape proposal related to the practical reuse of quarries and waste heaps, located in the Estremoz Anticline, will be shown.

Section 5.2 gives an overview about the geological and geomorphological aspects of the Estremoz Anticline. Section 5.3. formulates a systematic architectural methodology for the reuse of quarries and waste heaps. Section 5.4. shows the replanning of the industrial abandoned area of the Estremoz Anticline. Section 5.5. describes the process carried out for the selection of a specific intervention subarea, where to subsequently apply the architectural and landscape proposal. This process starts from the classification of the risk condition, to which the quarries and waste heaps are subjected, defining priority interventions, to a survey set up with a drone for the characterization of the selected quarry and waste heap. Section 5.6. illustrates the architectural and landscape proposal designed for a specific subarea. Finally, Sect. 5.7. draws the main conclusions.

Fig. 5.8 *Escombreiras* in the Estremoz Anticline. (Katia Talento, 2021)

5.2 Geological and Geomorphological Framework

5.2.1 Localization and Generalities

J. C. Kullberg
NOVA, Faculty of Sciences and Technology, Earth Sciences Department and GeoBiotec, University of Lisbon, Caparica, Portugal
e-mail: jck@fct.unl.pt

In what refers the tectono-stratigraphic units, in which the Portuguese continental territory is divided, the Estremoz Anticline is placed in the Ossa-Morena Zone (OMZ), an area where the geological units are of Neoproterozoic to Paleozoic age (Fig. 5.9). It is geographically located in the Portuguese province of Alentejo, in the Évora district.

Its shape is approximately elliptic with the longest axis-oriented NW-SE, measuring about 42 km by 8.5 km of maximum size in the shorter axis. The Estremoz Anticline spans from the Sousel (in NW end) to the Alandroal villages (in the SE end) being the most important extractive areas located in between, in the vicinity of the Estremoz, Borba, and Vila Viçosa villages (see, respectively, a), b), and c) in Fig. 5.9), forming an elongated triangle, the so-called the marble triangle, known by its industrial and commercial importance.

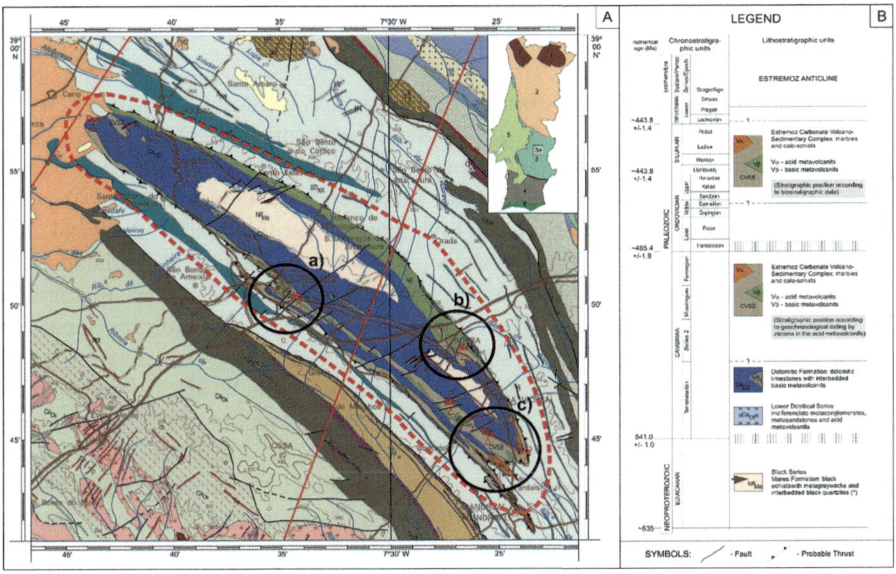

Fig. 5.9 Geologic map of the Estremoz Anticline (**a**) and the respective legend (**b**). The inset in (**a**) shows the general location (3a) of the represented area and the tectono-stratigraphic units of the Portuguese Continent: 1- Allochthonous Terrains (Bragança and Vinhais); 2- Centro-Iberian Zone; 3- Ossa Morena Zone; 4- South Portuguese Zone; 5- Western Margin; and 6- Algarve Margin. The red dotted area limits the "Estremoz Anticline" and the black circles highlights the main extraction areas of the "Marble Triangle": (a) the Estremoz nucleus, (b) the Borba nucleus, and (c) the Vila Viçosa nucleus. Extract from the Portuguese Geological Map at the 1/200.000 scale (295)

Thematically, the Estremoz Anticline is a distinctive feature represented in the Portuguese geological map at different scales, from the generic 1/1.000.000 to the 1/50.000 scale, being the more detailed one represented in two sheets, the 36D-Redondo e 36B-Estremoz sheets, published by the Laboratório Nacional de Energia e Geologia (LNEG), the former Portuguese Geological Survey. Due to the economic importance of this geological occurrence, Moreira and Vintém (Moreira & Vintém, 1997) published a more detailed map, at the 1/25.000 scale, with a higher definition of the lithostratigraphic units and a more expressive representation of a NE-SW to ENE-WSW fault system, which affect the Estremoz Anticline and corresponds to a direction with a regional geological significance.

5.2.2 Geometry, Tectonics, and Stratigraphy

The Estremoz Anticline is cartographically a slightly accurate structure with the convex side pointing to NE, and it is composed of several lithostratigraphic units, the older ones cropping out in the core of the anticline, meaning that geometrically and genetically the structure is an antiformal anticline more or less symmetrical in what refers to the bed dipping. We can describe those lithostratigraphic units, where the "Estremoz Marbles" are integrated, from the bottom to the top, as follows (Fig. 5.9):

- The older unit is the Mares Formation ("Mares Fm.," NPMa Fm.) which belongs to the Black Series ("*Série Negra*"), and it is formed by black schists with metagreywacke and interbedded black quartzites (of Edicarian age— ~ 635 to 541.0 ± 1.0 Ma—in the 1/200.000 scale map) (cf. e.g., (Gonçalves, 1970).
- The "Lower Detritical Series" (CaDt Fm.) Lower Detritical Series, the basal unit, in some places with a lenticular shape, is composed of undifferentiated meta conglomerates, metasandstones, and acid metavolcanics which mark the base of the Portuguese Phanerozoic, placed above an unconformity with the Ediacaran Fm. (cf. e.g. (Gonçalves & Carvalhosa, 1994).
- The "Dolomitic Formation" (CaCD Fm., already on the basis of the Cambrian period) is formed by dolomitic limestones with interlayered basic metavolcanic rocks (Vβ1 in the 1/200.000 map).
- The "Estremoz Carbonate Volcano-Sedimentary Complex" (CVSE Fm.—"Complexo Vulcano-Sedimentar Carbonatado de Estremoz" in the 1/200.000 map) is the lithostratigraphic unit where the "Estremoz Marbles" are integrated. As will be seen ahead, this is a very differentiated unit with intercalations of both acid and basic metavolcanic rocks (Vα and Vβ, in the 1/200.000 map, respectively), evidencing complex paleotectonic and paleoenvironmental contexts.

Although the general stratigraphic sequence of all units in the core of the anticline is undeniable, since the Estremoz Anticline is a normal polarity tectonic structure, the chronostratigraphic dating of the upper unit, the SVSE, has long been under discussion and a subject of strong scientific disagreement. This is due to the rarity of the fossil record, the position of the outcrops where they were collected in the general sequence, as well as the lack of continuity of the outcrops in the field.

After Araújo et al. (Araújo et al., 2013), the carbonates (the unit where the marbles belong) were traditionally considered of Cambrian age (Gonçalves, 1974), but after Perdigão (Perdigão, 1976) and Oliveira (Oliveira, 1984), they were attributed to the Ordovician or to the Silurian periods. More recently, works by Piçarra and Le Menn (Piçarra & Le Meen, 1994) and Piçarra and Sarmiento (Piçarra & Sarmiento, 2006) showed that at least the upper part of the marbles can be of more recent ages, namely the Upper Silurian or the Devonian.

More recently, based on isotopic dating after zircon sampling in interbedded acid metavolcanic rocks (the Vα member of the SVSE Fm.) (Pereira et al., 2012), a numerical age of 499 Ma was obtained (approximately the limit between the Miaolingian and the Furongian epochs, after the more up-to-date International Chronostratigraphic Chart published by the ICS/UIGS (K. M. Cohen et al., 2013)), thus favoring the lower position in the unit as depicted in Fig. 5.9.

In the very recent synthesis of the Iberian Geology published in five volumes (Piçarra et al., 2014; Quesada & Oliveira, 2019), it is considered that the "finding of early Devonian free-living tentaculitoids and crinoid columnal, occurring in marbles belonging to the 'Volcano-Sedimentary Complex of Estremoz,' opens the possibility of a complete reconsideration of the age of one of the most distinctive units in the Paleozoic basement of southwestern Iberia (Gutiérrez-Marco et al., 2019).

The protolith of the marbles before the regional metamorphism occurred are platform limestones which show evidence of subaerial exposure during the sedimentation. In fact, the irregularity of the upper layers, which are filled by black schists, and the differences in thickness between layers located inside and contouring these structures, allowed to conclude that those limestones were eroded originating from karstic landforms covered by clays rich in organic matter (L. Lopes, 2020). All these conclusions are very important to better understand the variety of marbles that occur in the Estremoz Anticline, as well as the geographic location in an intertropical latitude of the Ossa-Morena tectonic Terrane during the evolution of the carbonate platform, to better constrain paleotectonic and paleoenvironmental reconstructions.

In what concerns the tectonic context involved in the formation of the Estremoz Anticline, after Pereira et al. (Pereira et al., 2012), it is of consensus that it is a major Variscan structure of the Ossa-Morena Zone in Portugal, which results from the interference of two folding phases (Lopes, 2003). Therefore, the stratigraphic sequence was first deformed and metamorphized (under the conditions of the green-schist facies) during the Variscan orogeny (Carboniferous) producing folds with axial plane-oriented E-W, slightly dipping to the South, which implies a Northward vergence. The second phase showed a more brittle behavior, with the generation of cleavage of axial plane, but with a rotation of the general tectonic vergence to the NE, forming interference patterns. These interference patterns, observable both at the outcrop (quarries fronts) and the cartographic scales, are frequently responsible for the variation of the quality of the marbles, with evidence economic consequences at the level of exploration planning and sometimes the low the percentage of recovery of blocks with economic value as ornamental rock (see Lopes (2020) for a detailed explanation).

The difficulties to achieve a consensual dating for the CVSE Fm. and the different interpretations that exists, related to the nature of contact between the lithostratigraphic units of the Estremoz Anticline and the involving ones (of Silurian age), also imply difficulties for the structural model. In fact, the authors who defend a more recent dating (Upper Silurian-Devonian age) for the CVSE Fm. propose a thrust contact between both series, meaning that the Estremoz Anticline must be a tectonic window of allochthonous units (the involving ones) thrusted over the autochthonous units of the Estremoz Anticline series. That is the meaning of the "Probable Thrust" depicted in turn of the core of the Estremoz Anticline in the 1/200.000 geological map partially represented in Fig. 5.9.

5.2.3 *Marble Varieties in the Estremoz Anticline*

The variety and complexity of the environmental conditions, related to the formation of the original limestones and the complexity of the tectonic evolution of the sector, are responsible for a wide variety of marbles that are explored in the "Estremoz-Borba-Vila Viçosa marble triangle" (LNEG, 2020) (Geoportal, n.d.).

Consulting the Catalogue of Portuguese Ornamental Stones (Moura & Martins, 1992), one can find the detailed technical description of 35 varieties of marble. This catalog is mainly destined to the building sector, namely for architects and

Fig. 5.10 Some of the most representative varieties of marbles extracted from the "Marble Triangle." From left to right "Ruivina from Lagoa," "Tiger skin from Pardais," "Pink with stripes from Borba," "Cream from Pardais," "Pink Dawn," and "Cream from Pard. (Modified from *M*. Lopes & Martins, 2015)

engineers, and the geological framework which conditioned this variety was not so important for the selection of the different types. These 35 varieties are extremely asymmetrically distributed by the three-extraction nucleus, as we can see: 3 in the Estremoz nucleus (located in the long limb of the anticline, in Figure 5.9a), 12 in the Borba nucleus (located in the short limb of the anticline, in Fig. 5.9b) and 20 in the Vila Viçosa nucleus (mainly located in the SE hinge of the anticline, in Fig. 5.9c).

On the other hand, the CEVALOR's Catalogue of Ornamental Stones (CEVALOR, undated) considerably reduces the varieties to a total of 17, without any concern for their distribution by the three nucleuses. Apparently, some of the varieties described are extracted from more than one nucleus.

Lopes and Martins (Lopes & Martins, 2015) in a synthesis work related to the "Estremoz Marble" (*lato sensus*) as a Global Heritage Stone, although still incomplete, built the basis for a simpler classification, emphasizing both color and textural characteristics (Fig. 5.10).

5.3 Systematic Architectural Methodology for the Rehabilitation of Quarries and the Reuse of Waste Heaps

The current research, starting from a theoretical review about the concept of landscape mutation, quarry rehabilitation and waste reuse, investigates new rehabilitation strategies for the industrial area of the Estremoz Anticline, by reusing the waste piles from the extraction activity. The general objective of the investigation is to re-qualify the abandoned industrial landscape using volumes and materials available in situ (quarries and waste mounds), to preserve the identity of the place, along with its historical and industrial memory.

Innovative strategies are proposed to demonstrate the possibility of transforming a disused industrial landscape into an attractive and visited one, having an associated aesthetic value.

In accordance with the principles previously elaborated, the research proposed in this book is mainly focused on re-establishing a recreational, cultural, naturalistic, productive, and touristic purpose, replanning the area, revitalizing the abandoned landscape with new functions, and considering the quarries and the waste heaps as monumental "containers" of spaces, integrated in the landscape, and communicating with the pre-existence. The innovation of this research also consists of involving the waste piles in the process of rehabilitation of the abandoned extractive site, integrating them into the landscape.

The practical analysis of the case studies (related to the rehabilitation of quarries and the reuse of waste heaps) and the theoretical understanding about circular economy, discussed in previous chapters, have jointly formed the basis of the research to comprehend how the idea of reusing can advance in the architectural discipline, following a new systematic methodology (as it will be shown in this section). The work adsorbs the main concepts, described in the preceding sections, to materially deal with the problem of the mounds of waste from quarries. These foremost concepts can be shortened as follows:

I. The beauty corresponds to the essence of a thing and not to the idea of its form (De Poli & Incerti, 2014).
II. The rehabilitation process should protect the traces of the industrial activity, as testimonies of the history of the place (Talento et al., 2019).
III. The reuse is an indispensable component to understand how to reprocess an obsolete matter into something that is put back into cycles of use (Wong, 1986).
IV. The "*superuse*" is a concept that differs from simple "reuse" for the need to propose future scenarios for the rejected material (Use It More. Superuse Studios, n.d.).
V. It is necessary to create different business opportunities "discovering" local and pre-used materials and not to use new and innovative technologies (D. Cohen, 2007).
VI. It is challenging to act in situ, not exporting the waste material.
VII. One must consider the giant waste heaps as monumental components, integrated into the landscape and interactive with people.

Based on this premise, the current section proposes the development of a systematic architectural methodology, never experienced and applied before (to the best of the author's knowledge), to deepen the waste question in the architectural context, considering the quarries and the waste piles as principal subjects of the rehabilitation process of an extractive area. For this purpose, it has been considered important to make an analysis of the study area, the Portuguese Estremoz Anticline, and to identify it on territorial maps. Through this identification, one realizes the vast dimensions of the area (about 280 km^2) and the significant number of pits and waste piles located on it (about 476).

Before replanning the area, useful for the rehabilitation of the Estremoz Anticline, it is essential to create a scientific formula (as it will be presented in this section) to confer renovated functions, from a landscape and architectural point of view, to the abandoned industrial space. The elaboration of this original formula first involves the characterization of the quarries and waste piles located in the Estremoz Anticline, through the allocation of metadata to each of them, which can be divided into two main categories:

I. **ArcGIS Key Parameter Inputs ("Static Conditions")**

- Geographical coordinates.
- Roads and streets maps.
- Aquifer map.
- Geological map.
- Current Masterplan Plano Diretor Municipal (PDM) in Portuguese and corresponding mapping.

II. **Features ("Variables," listed in order of priority— $\vec{F}$)**

- *State of activity* (active, inactive, abandoned, or suspended): To understand in which quarry it is possible to intervene. For the active quarries, only minimal interventions are permitted.
- *Dimensions* (area, volume, maximum depth, and height): To know the physical and structural features of the quarries.
- Density: The number of quarries and *escombreiras*, located in a specific fraction of the general area, affects the type of use to be allocated to that area.
- *Visual impact*: To analyze how the waste piles are viewed from the main streets and urban center. This data is relevant to improve the aesthetic aspect of the accumulations.
- *Proximity to urban center* (<1 km): to define the most appropriate destinations of use of the quarries and *escombreiras*, which should be coherent with the activity of the urban center.
- *Identification of the context* (urban, semi-rural and rural): To select the specific use of the quarries and *escombreiras* (appropriate to the area) during the replanning operation.
- Accessibility: This criterion is crucial in the replanning process to understand if the touristic, cultural, educational, recreational, naturalistic, and productive structures are easily accessible by residents and tourists.
- Proximity to attractions.
- Proximity to an economic activity.
- Proximity to a sport activity.
- Proximity to a cultural activity.
- Peculiar morphology.

As explained in the previous chapter, through the literature review about the concept of quarry rehabilitation and waste reuse, this research, starting from the analysis of several case studies, has formulated important typologies of reuse for the quarries and the accumulation of wastes. These typologies are resumed in Table 5.1.

Associating the rehabilitation and reuse typologies with the features of each quarry and *escombreira*, the research creates a new formula to define new destinations of use to confer them:

$$[\boldsymbol{D}_U] = \overrightarrow{\boldsymbol{T}_R} \times \vec{\boldsymbol{F}}$$

$[\boldsymbol{D}_U]$ = Destination of use (matrix)
$\overrightarrow{\boldsymbol{T}_R}$ = Typology of rehabilitation/reuse (vector)
$\vec{\boldsymbol{F}}$ = Feature (vector)

Through this formula of practical use, the work, presented in this book, creates the primary criteria of regeneration and the main guidelines for planners to understand which type of reuse to adopt, in specific situations, in any part of the world (Figs. 5.11 and 5.12).

The elaboration of this original formula, jointly with the rehabilitation and reuse typologies for quarries and waste heaps, and the primary guidelines for planners, embodies the systematic methodology so far discussed, and it will be extremely crucial for the replanning of the industrial area of the Estremoz Anticline, as it will be shown in the next section.

5.4 Architecture and Landscape: Replanning the Estremoz Anticline

5.4.1 The "Eating an Elephant" Theory for the Landscape Redesign

As defined before, the general objective of the current research is to re-qualify the industrial landscape of the Estremoz Anticline, using volumes and materials available in situ (quarries and waste mounds), to preserve the historical and industrial memory of the place and to demonstrate the possibility of transforming a disused landscape into an attractive and visited one, having an associated aesthetic value. This general objective includes the transformation of the abandoned quarries and waste heaps in "containers of spaces" through a design of an innovative architectural and landscape proposal. Being the Estremoz Anticline, a massive area (about of 280 km^2), and containing a huge number of quarries and wastes, it seems obvious that, in the context of the current research, it is not possible to design a proposal for the entire area. For this reason, there is a question that arises spontaneously:

Table 5.1 Resume of different typologies of rehabilitation for quarries and reuse for waste heaps ($\overrightarrow{T_R}$). (Katia Talento, 2021)

	Quarries	Waste heaps	
Typology of rehabilitation	Discussion	Typology of reuse	Discussion
Filling	Operation that involves the use of water, ground, or constructions to fill the open pit	***Detachment***	Insertion of small ephemeral structures in proximity to the waste piles, not interfering with them. There are respect and a valorization of the pre-existence by contemplating it
Modification	*Method that includes the change of the form to regenerate the quarry, the adoption of actions of sculptural remodeling (emphasizing) or to hide undesired elements through technological solutions (camouflage)*	***Equality***	It consists in treating the accumulation of scraps as the other components of the landscape. The waste accumulation represents not an isolated factor to monumentalizing but a territorial component able to interact with people
Insertion	*It comprises interventions of new constructions partially inserted in the volume of the quarry, causing excavations*	***Binomial monumentality-human interaction***	It includes both monumentality and the interaction components, exalting the waste accumulation as a prominent installation, opening it at people visits
Small interventions	Method used for all the places, which, although deformed, present special features of form. For all these cases, only small interventions are recommended, related to the securing of the sites and the excavation	***Naturalistic***	Mimesis with the landscape. The waste heap is treated as an integral part of the landscape and not as an extraneous object to contemplate. This criterion allows the introduction of recreational and cultural structures only if they appear "hidden" in the waste accumulation
Attached	Measures that adopt the construction of buildings, paths, and facilities to resolve the metamorphosis of the quarry	***Architectonic***	It completely exalts the meaning of the pile of remains giving it the function of "shell." the waste pile can contain buildings and renewed spaces
Natural interventions	They exalt the power of nature to destroy any form of degradation. Nature will restore the lost balances through spontaneous biological processes		

Reuse typology	Features	State of Activity: Active	State of Activity: Inactive	State of Activity: Suspended/ Abandoned	Quarry type: Mount	Quarry type: Plain	Dimensions: Small	Dimensions: Medium	Dimensions: Large	Density: Low	Density: High	Visual Impact: Low	Visual Impact: High	Proximity to urban centre (<1km)	Context: Urban	Context: Semi-rural	Context: Rural	Accessibility: Favourable	Accessibility: Not Favourable	Proximity to attractions	Proximity to economic activity	Proximity to sport activity	Proximity to cultural activity	Planning result	Architectural result
Filling	Water		•	•		•		•	•		•	•				•	•	•						Naturalistic and recreational area	Natural parks with lakes
			•	•		•		•	•	•		•				•	•	•						Agricultural and naturalistic area 1	Irrigation reservoir
			•	•		•	•	•		•		•		•	•			•				•		Sports area	Swimming pools
	Ground/ Inert		•	•		•	•			•		•				•	•	•						Naturalistic and Agricultural area 2	Revegetation
	Building		•	•		•	•	•		•		•		•	•			•		•	•	•	•	Economic, cultural and sports area	Touristic structures, accomodations, buildings for shopping, wineries, food courts, cultural and sports structures
Modification	New Form		•	•		•	•	•	•	•		•				•	•	•						Naturalistic, agricultural, and recreational area	Revegetation and natural parks
	Emphasizing	•	•	•	•	•		•	•		•		•		•	•	•	•		•				Industrial interpretational area	Minimal interventions: securing, paths, rest stops and info points.
Insertion	New construction		•	•	•	•	•	•	•	•			•	•	•			•		•			•	Cultural area	Museums, exhibition spaces, galleries, theatres
			•	•	•	•	•	•	•	•			•	•	•			•			•			Economic area	Infrastructures, Touristic buildings, accomodations, buildings for shopping, wineries, food courts
			•	•	•	•	•	•	•	•			•	•	•			•				•		Sports area	Sports structures
			•	•	•	•	•	•	•	•			•			•	•	•						Economic area	Rural tourism, accommodations, wineries
Small interventions		•			•	•		•	•		•	•	•		•	•	•	•						Industrial interpretational area	Minimal interventions: securing, paths, rest stops and info points.
Attached	Horizontally		•	•	•			•	•		•		•			•	•	•						Economic area	Touristic structures
	Vertically		•	•		•	•	•	•	•		•		•	•			•		•	•		•	Economic and cultural area	Touristic structures, accomodations, buildings for shopping, wineries, food courts, cultural structures
	Suspended		•	•	•			•	•		•		•			•	•	•						Economic area	Touristic structures
Natural interventions			•	•	•	•	•	•	•	•	•	•	•	•	•	•	•	•	•	•	•	•	•	Naturalistic area	Revegetation and biodiversity recovery

Fig. 5.11 Primary guidelines for the rehabilitation of quarries [$\boldsymbol{D_U}$]. (Katia Talento, 2021)

PRIORITY →

Features / Reuse typology	State of activity: Active	State of activity: Inactive	State of activity: Suspended/ Abandoned	Dimensions: Small	Dimensions: Medium	Dimensions: Large	Density: Low	Density: High	Visual impact: Low	Visual impact: High	Proximity to urban centre (<1km)	Context: Urban	Context: Semirural	Context: Rural	Accessibility: Favourable	Accessibility: Not favourable	Proximity to attractions	Proximity to economi c activity	Proximity to sport activity	Proximity to cultural activity	Peculiar morphology	Planning result	Architectural result
Detachment	•				•	•		•		•	•	•			•		•					Industrial interpretational area	Minimal interventions: securing, paths, rest stops and info points.
	•				•	•		•		•			•	•	•							Industrial interpretational area	Minimal interventions: securing, paths, rest stops and info points.
Equality		•	•		•	•	•			•	•	•			•				•			Sports area	Playground
Binomial monumentality-human interaction		•	•		•	•	•			•	•	•			•		•			•	•	Cultural and recreational area	Open air museums, exhibition spaces and terraces
Naturalistic		•	•		•	•		•		•			•	•	•							Agricultural and naturalistic area 1	Cultivated terraced gardens
		•	•	•			•	•	•				•	•	•							Naturalistic and recreational area	Natural parks
		•	•	•			•		•				•	•	•	•						Agricultural and naturalistic area 2	Revegetation
Architectonic		•	•		•		•			•	•	•			•			•				Economic area	Touristic structures, accomodations, buildings for shopping, wineries, food courts
		•	•		•		•			•	•	•			•		•			•		Cultural area	Museums, exhibition spaces, galleries, and cinema
		•	•		•		•			•	•	•			•				•			Sports area	Sports structures

Fig. 5.12 Primary guidelines for the reuse of waste heaps [$\boldsymbol{D_U}$]. (Katia Talento, 2021)

Which quarries and waste heaps will be involved in the architectural and landscape design process and how will they be selected?

After having physically individuated the quarries and the mounds of wastes on the territorial map, and after having elaborated a scientific formula for the definition of new destinations of use, it is important to introduce a component of land-use planning, considering landscape units for the area under investigation. For this effect, the research thinks on the macro scale through an urban replanning of the Estremoz Anticline, imagining new macro-uses for the abandoned areas of the extractive industry that they do not conflict with all those where there are still active quarries. There is an old South African proverb that says:

When you want to eat an elephant that's in your way, make it slice by slice. [Desmond Tutu]

In other words, the best way to solve a big problem is to tackle it a little at time. If the elephant metaphorically represents the huge landscape, one will have to understand it, give it identity, respect it in the ecosystem, understand its habits and language, and (very important) know how one perceives it and makes it belong to "our world." Thus, characterized, understood, and considering it as a "partner" in our life, one knows this landscape and, knowing also where it came from (how one slices it), one perceives the present to give an opinion about the future. At this stage, one knows the elephant and what one wants from him, so the three major steps of the proposal will be defined:

I. *"Let's eat the elephant into slices"*: It means to create the segmentation of the landscape of the Estremoz Anticline.
II. *"After having cut the elephant and identified all its parts, the piece to be immediately eaten was chosen"*: It refers to define the scale of intervention, which will not necessarily correspond to the entire universe of quarries. It might only be an illustrative subsegment of the research theory.
III. *"All the rest of the animal will be eaten later"*: It implies the preservation of the remaining quarries for future developments.

This analogical metaphor between the elephant and the huge landscape of the Estremoz Anticline analyses what is intended by the research: to identify, perceive, and respect the original industrial landscape, to unravel its details, make a global management proposal, and model a design of a small excerpt. Based on this premise, a planning operation of the extractive area has been developed. The extractive territory to regenerate includes the municipalities of Estremoz, Borba, and Vila Viçosa, for which a different strategy of rehabilitation has been adopted, based on the diverse features they have. The zoning of the extractive area is described in detail in the following subsections:

5.4.2 *Estremoz*

Most quarries belonging to this municipality are located inside the original area destined to the extraction of marble, respecting the current Masterplan (PDM). In this part of the territory, the planning pays particular attention to the environmental and natural conditions of the surrounding to integrate the industrial abandoned zone in this specific rural landscape. For this reason, two typologies of natural landscape have been designed to redevelop the area:

I. **Natural and agricultural zone**: Due to the minimal dimensions of the quarries and *escombreiras* and the spontaneous development of a moderate or high quantity of flora inside and above them, a complete revegetation of the area has been planned to recreate the original landscape for natural and agricultural purposes (Fig. 5.13).

II. **Natural and agricultural zone with terraced gardens**: It consists in recreating the natural conditions of the context, promoting the agriculture on the *escombreiras*. In this way, the mounds of waste, having consistent dimensions, actively participate in the process of planning. Contrary to what happens in the north of Portugal, in the Douro Valley, the presence of terraced gardens and fields is not common in the Alentejo, where the agriculture is practiced on flat lands. Since the municipality of Estremoz (as well as the other towns belong to the Anticline) is strongly composed by scrap accumulations that cannot be demolished or dis-

Fig. 5.13 Planning of the industrial area of the municipality of Estremoz. (Katia Talento, 2021)

posed elsewhere due to economic reasons, the research proposes a new way to develop the local agriculture. The consequence of this solution is not only the progress of this activity but also an aesthetic progress, capable of conferring a better image of the abandoned industrial area. For this reason, the *escombreiras* will become green and more valuable elements, integrated in the landscape, with a lower visual impact, and the associated quarries will turn into reservoirs for regulating and storing irrigation water (Fig. 5.13).

Additional uses are related to:

III. **Industrial interpretation zone**: It respects the industrial activity that is still ongoing. To promote the marble extraction, the planning process does not include significant interventions for this area. There are respect and a valorization of the industrial activity by contemplating it (detachment typology of reuse). For this purpose, only minimal interventions will be considered, such as securing and the introduction of paths, rest stops, and info points to also confer a cultural and an educational value to the area (Fig. 5.13).

IV. **Cultural zone**: It consists in conceiving a space for events, cultural and entertainment activities. This zone arises from the need to confer a more appropriate use to the quarries and *escombreiras* located very close to the urban center and a cultural area, which has already been planned through the current PDM. For this situation, the quarries and *escombreiras* are considered as "containers" of spaces, related to the creation of museums, amphitheaters, and exhibitions spaces, exalting the monumentality of the pre-existence and, at the same time, promoting human interactions (Fig. 5.13).

Not all the quarries are conforming to the Portuguese PDM. A significant amount of them is located outside the area destined to the extraction activity. For these quarries, the planning operation has considered the most suitable use, coherent with the surrounding. Based on these reflections, two additional areas have been redesigned in the municipality of Estremoz:

V. **Sport and recreational zone**: It derives from the presence of quarries and *escombreiras* placed in proximity to the residential area. The sport activity is promoted through open-air or indoor facilities, creating playgrounds, sport structures, and swimming pools (Fig. 5.13).

VI. **Economic zone**: The economic use has been conferred to those quarries and *escombreiras* that are located inside the economic area defined by the Portuguese PDM. The concept of considering the *escombreiras* as principal elements of the rehabilitation of the industrial area will improve the touristic flow. For this reason, the planning includes accommodation facilities that are located inside the piles of remains with moderate dimensions. The associated quarries, with an operation of filling, will turn into leisure places, such as swimming pools and bathing area (Fig. 5.13).

5.4.3 Borba

The allocation of the quarries in this municipality occurs in two distinct areas; the first one is placed close to the urban center and the second one is more peripherical and integrated in the rural context. Besides, all the quarries respect the area destined to the extraction activity, illustrated in the current Masterplan (PDM). The destinations of use, planned for the municipality of Estremoz, have also been considered in this case, even with different reasons. In the first area (close to the urban center), the planning operation provides for the following zoning:

I. **Sport and recreational zone**: It is located in proximity to the existing sports equipment. Therefore, this portion of the abandoned area could be considered as the extension of the sport area already present on the territory, having also indoor facilities (Fig. 5.14).
II. **Industrial and interpretation zone**: In this area, there are both abandoned and active quarries. For this reason, this destination of use derives from the need to highlight and preserve the industrial memory of the place and, at the same time, to respect the extraction activity, which is still ongoing. Since the quarries, and the related *escombreiras*, have huge dimensions and they are subjected to risk conditions, only minimal interventions could be made to regenerate the area.

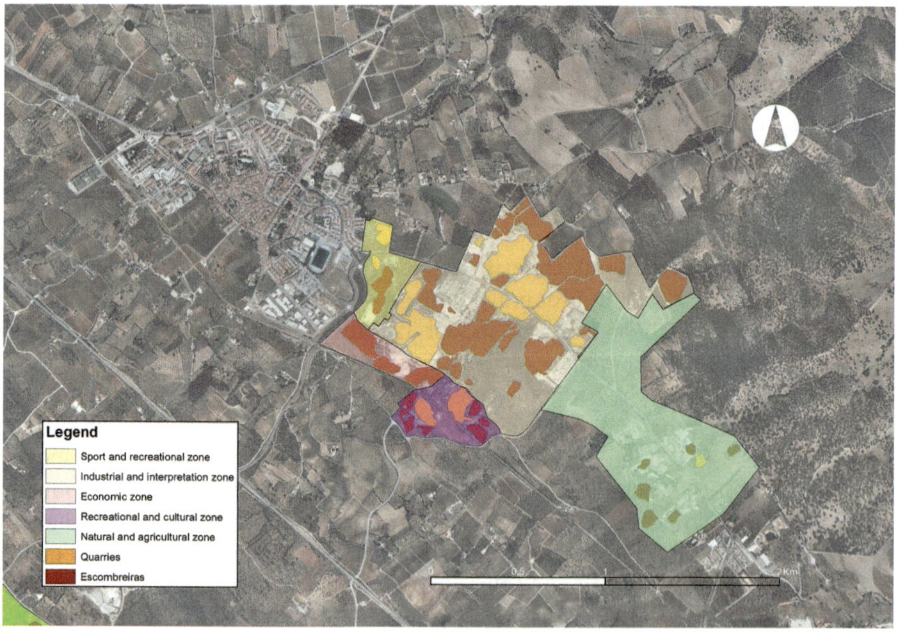

Fig. 5.14 Planning of the industrial area of the municipality of Borba: first area. (Katia Talento, 2021) In the second area, the peripherical one, integrated in the rural context, the natural landscape presents three interesting variations

Few examples are the securing of the quarries and *escombreiras* and the creation of paths, rest stops, and viewpoints to consider this area as an industrial park with cultural value (Fig. 5.14).

III. **Economic zone**: It represents an extension of the existing economic area, and it includes accommodations and touristic structures. Two typologies of reuse belong to this area; the "architectonic" one, if the buildings will be designed inside the mounds of wastes, and the "equality" one, if the structures will be considered as real mountains on which to realize mini touristic villages (Fig. 5.14).

IV. **Recreational and cultural zone**: Thanks to the peculiar morphology of the area and the analysis of the scientific criteria, previously formulated, this area appears adequate to the organization of cultural and recreational events, such as concerts, shows, exhibitions, and trade fairs (Fig. 5.14).

V. **Natural and agricultural zone**: It has designed to re-establish the original conditions of the landscape, thanks to the presence of a limited number of quarries and *escombreiras* with irrelevant dimensions (Fig. 5.14).

VI. **Natural and agricultural zone**: Due to the minimal dimensions of the quarries and *escombreiras*, the operation of revegetation to recreate the original condition of the landscape has been proposed for this area (Fig. 5.15).

VII. **Natural and bathing zone**: It provides for the creation of an artificial system of lakes thanks to the presence of many flooded quarries with a shallow depth.

Fig. 5.15 Planning of the industrial area of the municipality of Borba: second area. (Katia Talento, 2021)

The *escombreiras* dimensions are minimal, and they could be perfectly reintegrated in the landscape through an intervention of revegetation (Fig. 5.15).

VIII. **Natural and agricultural zone with terraced gardens**: Due to the morphology of the area and the presence of many *escombreiras* with moderate and elevate dimensions, it is possible to introduce terraced spaces with agricultural productions and reservoirs for regulating and storing irrigation water (Fig. 5.15).

5.4.4 Vila Viçosa

Before proceeding with the urban planning of the last municipality, Vila Viçosa, it is important to confer a general overview about the industrial situation. Since Vila Viçosa represents the place where a specific intervention area has been selected for the practical application of the reuse topic, through the design of an architectural and landscape proposal, it is valuable to create a more intensive analysis. For this effect, all the parameters associated with the quarries and *escombreiras* will be presented. This analysis has also been elaborated for the previous municipalities but, to ensure the fluence of the text, it has been inserted within the Annex of the present research.

State of Activity

In accordance with what has been previously described (Sect. 5.2), the analysis starts with the identification and characterization of all the quarries and *escombreiras* parameters, belonging to Vila Viçosa, differentiating them first by state of activity. To identify the state of the activity of each quarry (Fig. 5.16) and *escombreira* (Fig. 5.17), reference has been made to the following terms (DGEG—Divisão de Pedreiras do Sul, 2016):

- **Active**: It refers to an operating quarry with an allocated right of exploitation.
- **Inactive**: It indicates an interruption, for a period of less than 2 years of unreported and unauthorized exploration and recovery.
- **Suspended**: It refers to an authorized interruption.
- **Abandoned**: It means an unauthorized interruption of exploration, for a period exceeding 2 years.

Dimensions

The analysis proceeds with the differentiation of the quarries and the *escombreiras* by dimensions (Figs. 5.18 and 5.19). For this reason, the following data has been considered:

Dimension	Area	Volume	Height
Extra small	<500 m^2	<1.000 m^3	<5 m
Small	500 m^2 < A < 1.000 m^2	1.000 m^3 < V < 10.000 m^3	<5 m

Dimension	Area	Volume	Height
Small<D < medium	1.000 m^2 < A < 5.000 m^2	10.000 m^3 < V < 20.000 m^3	5 m < h < 10 m
Medium	5.000 m^2 < A < 10.000 m^2	20.000 m^3 < V < 200.000 m^3	5 m < h < 10 m
Medium<D < large	10.000 m^2 < A < 20.000 m^2	20.000 m^3 < V < 200.000 m^3	10 m < h < 15 m
Large	20.000 m^2 < A < 50.000 m^2	200.000 m^3 < V < 500.000 m^3	15 m < h < 40 m
Very large	20.000 m^2 < A < 50.000 m^2	>500.000 m^3	>40 m
Very large	>50.000 m^2	>500.000 m^3	15 m < h < 40 m
Very large	>50.000 m^2	>500.000 m^3	>40 m

Dimension	Area	Depth
Extra small	<500 m^2	<5 m
Small	1.000 m^2 < A < 5.000 m^2	5 m < d < 10 m
Medium	5.000 m^2 < A < 10.000 m^2	5 m < h < 10 m
Medium<D < large	10.000 m^2 < A < 20.000 m^2	10 m < h < 15 m
Large	20.000 m^2 < A < 50.000 m^2	15 m < h < 40 m
Very large	20.000 m^2 < A < 50.000 m^2	>40 m
Very large	>50.000 m^2	15 m < h < 40 m
Very large	>50.000 m^2	>40 m

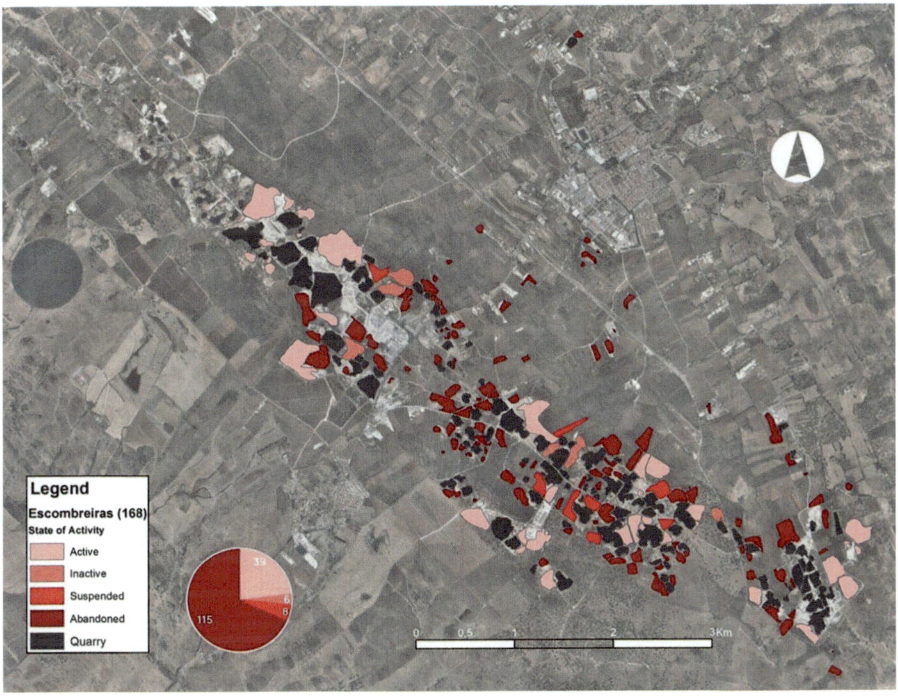

Fig. 5.16 *Escombreiras*' state of activity. (Katia Talento, 2021)

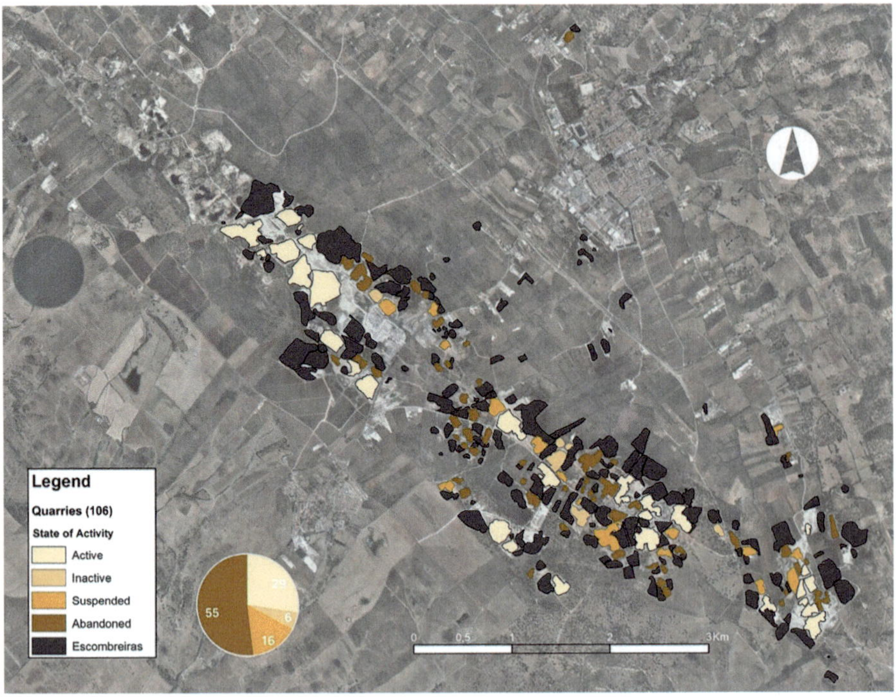

Fig. 5.17 Quarries' state of activity. (Katia Talento, 2021)

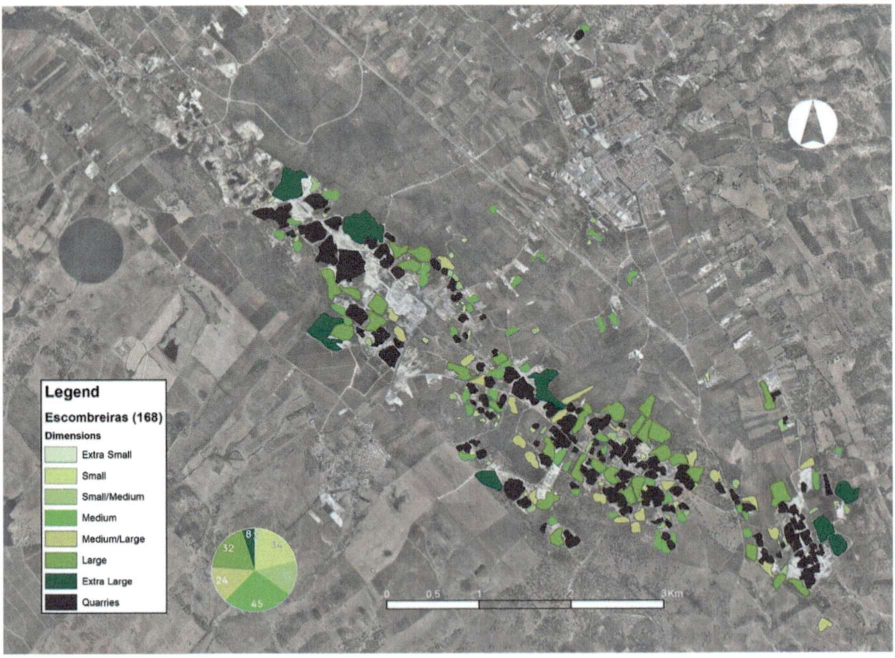

Fig. 5.18 *Escombreiras*' dimensions. (Katia Talento, 2021)

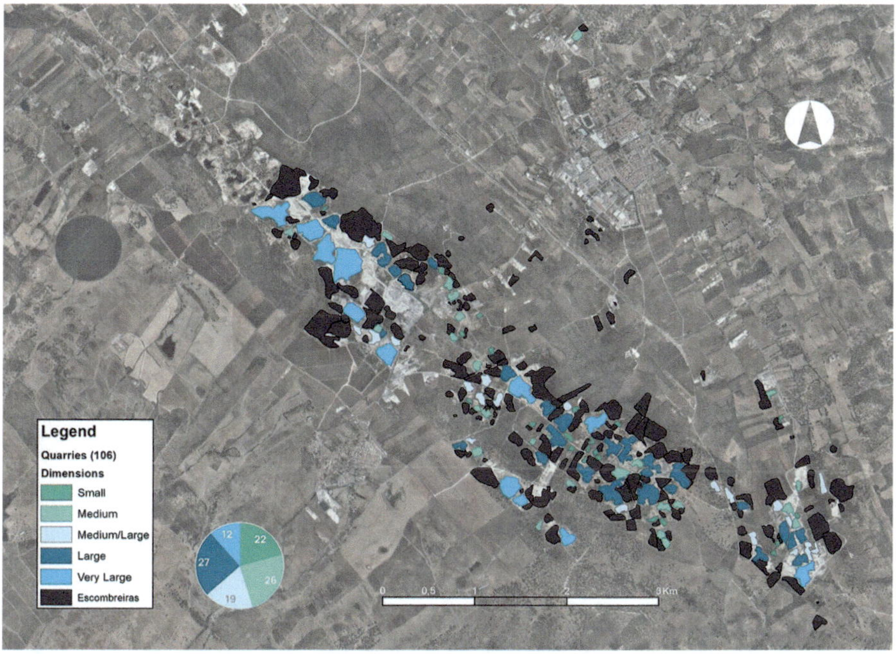

Fig. 5.19 Quarries' dimensions. (Katia Talento, 2021)

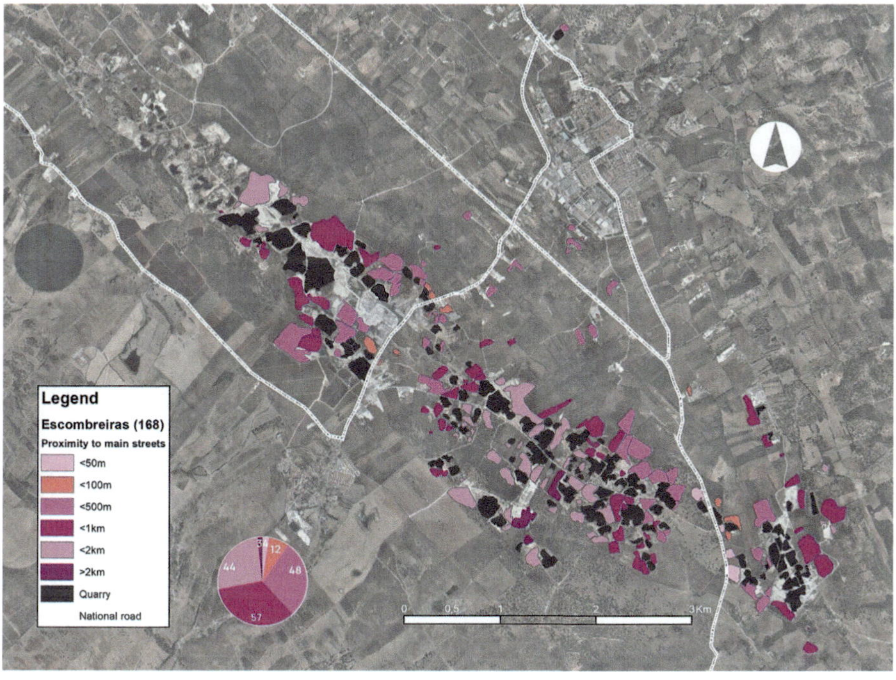

Fig. 5.20 *Escombreiras*' proximity to main streets. (Katia Talento, 2021)

Proximity to Main and Secondary Streets

This criterion is crucial in the replanning process to understand if the touristic, cultural, educational, recreational, naturalistic, and productive structures are easily accessible by residents and tourists. It is also important to understand which quarries and *escombreiras* respect the minimal distance from roads, required by law, to prevent risk conditions (Figs. 5.20, 5.21, 5.22, and 5.23).

Street	Distance
Main street	>50 m
Secondary street	>15 m

Proximity to the Urban Center

The analysis of this feature allows to define the most appropriate destinations of use of the quarries and *escombreiras*, which should be coherent with the activity of the urban center (Figs. 5.24 and 5.25).

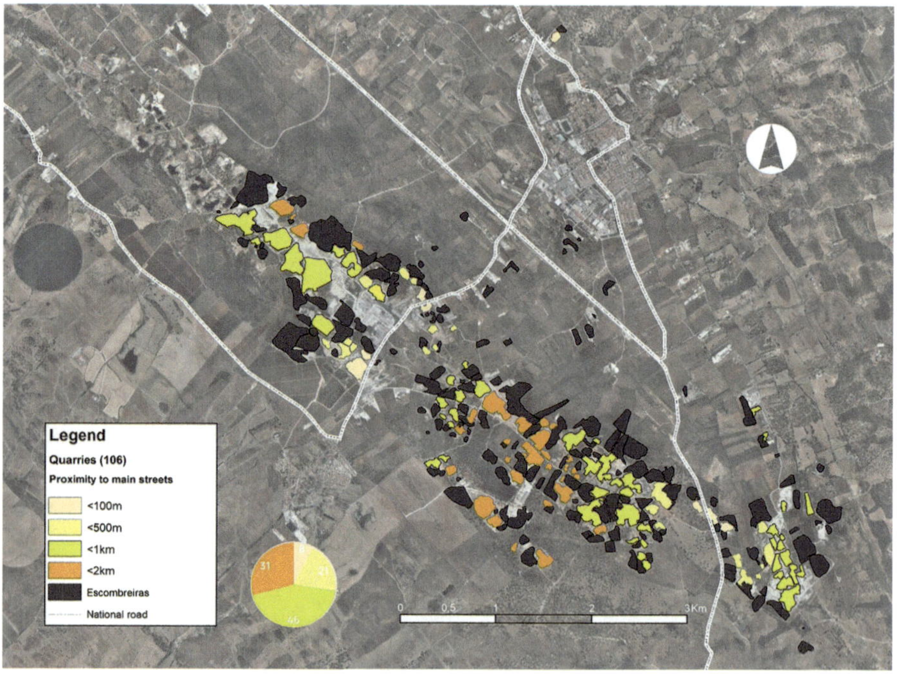

Fig. 5.21 Quarries' proximity to main streets. (Katia Talento, 2021)

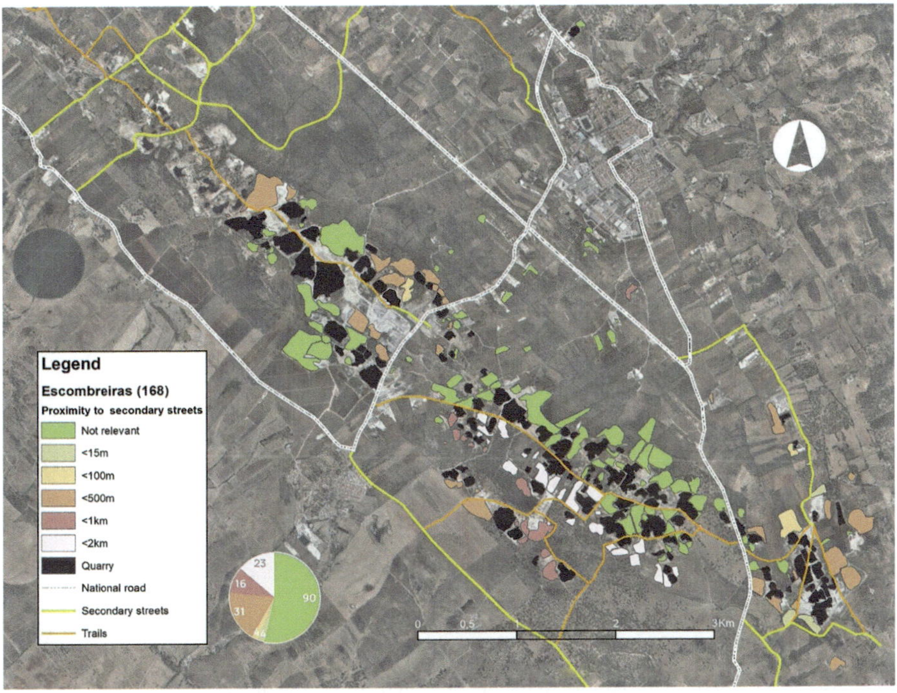

Fig. 5.22 *Escombreiras*' proximity to secondary streets. (Katia Talento, 2021)

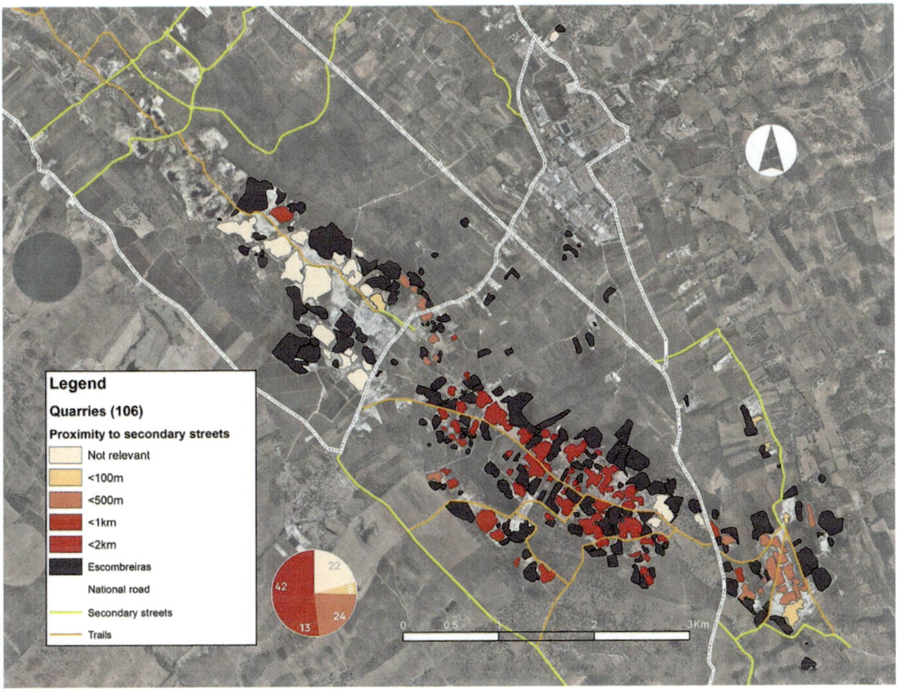

Fig. 5.23 Quarries' proximity to secondary streets. (Katia Talento, 2021)

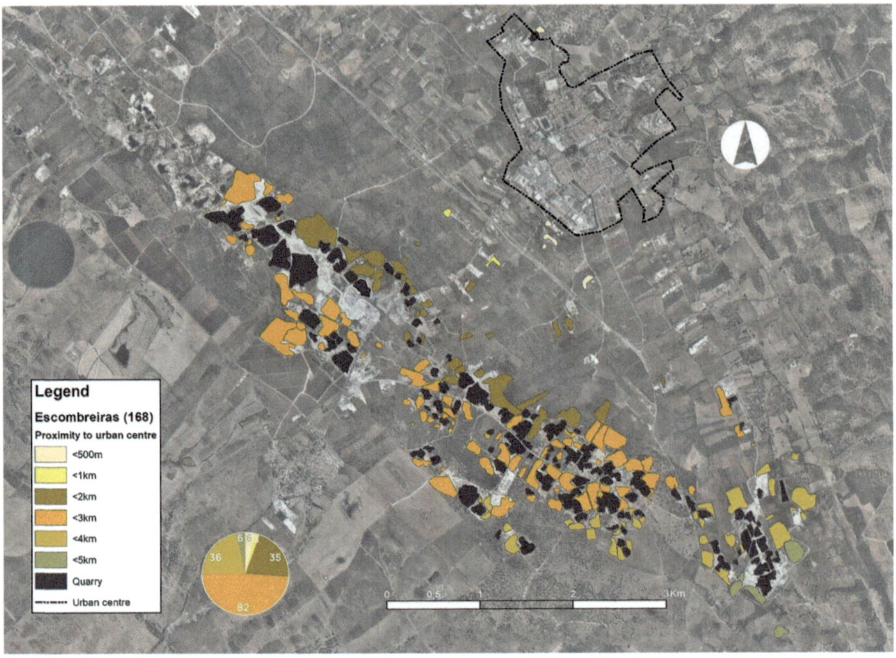

Fig. 5.24 *Escombreiras*' proximity to the urban center. (Katia Talento, 2021)

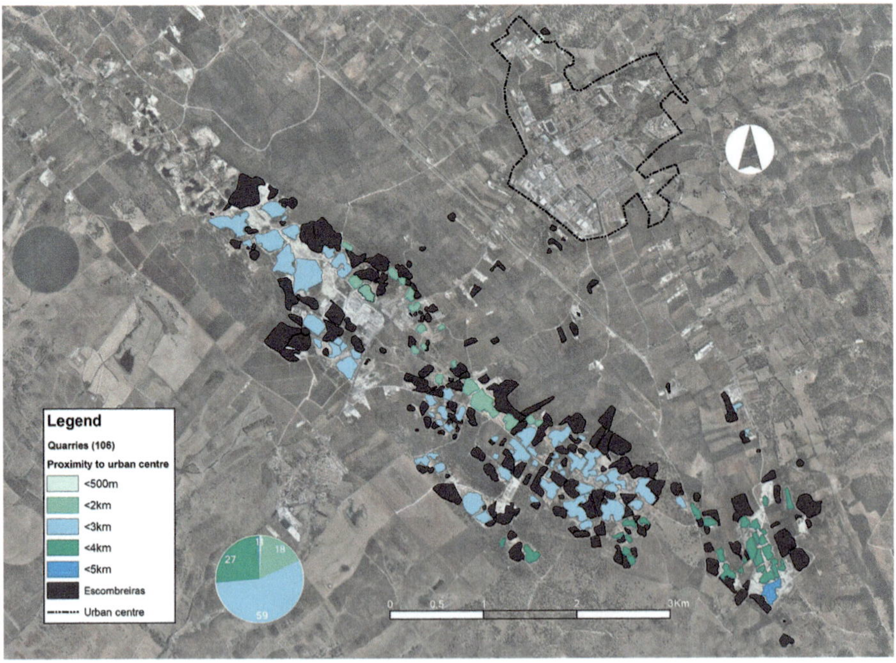

Fig. 5.25 Quarries' proximity to the urban center. (Katia Talento, 2021)

Context

This feature differentiates the quarries and the *escombreiras* depending on if they are in an urban, semi-rural, or rural context (Figs. 5.26 and 5.27). In this way, the most suitable use can be allocated during the replanning operation. The following data has been considered:

Context	Distance from urban center
Urban	<1.000 m^2
Semi-rural	<2.000 m^2
Rural	<5.000 m^2

Visual Impact

This data is useful to analyze how the *escombreiras* are viewed from the main streets and the urban center, and it is relevant to improve the aesthetic aspect associated to the accumulations of wastes (Fig. 5.28).

Main Attractions

The identification of the touristic attractions of the Municipality allows to define the most proper destinations of use for those quarries and *escombreiras* that are in proximity to them, valorizing the attraction itself and creating consistency with the context (Fig. 5.29).

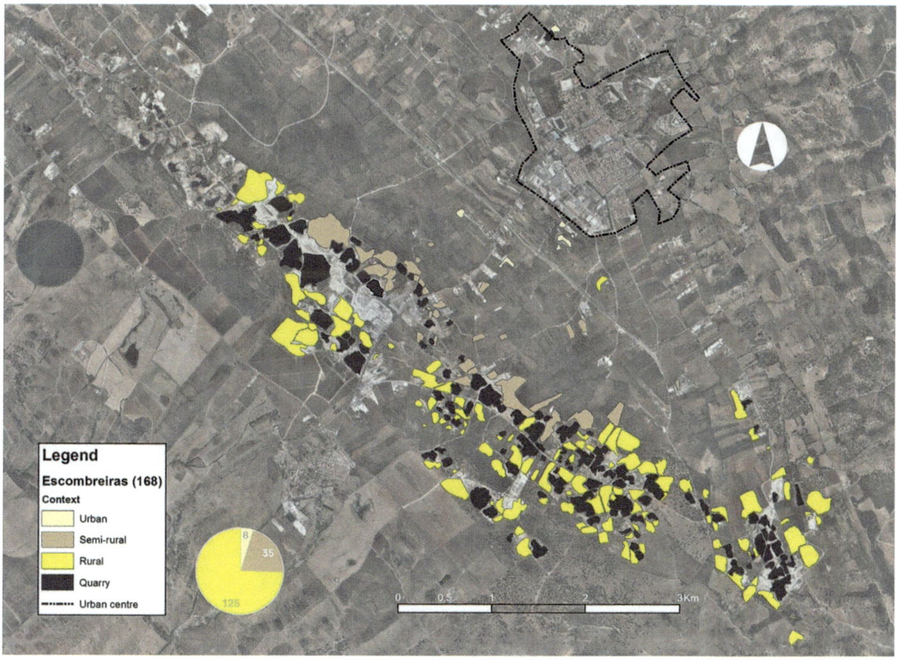

Fig. 5.26 *Escombreiras*' context. (Katia Talento, 2021)

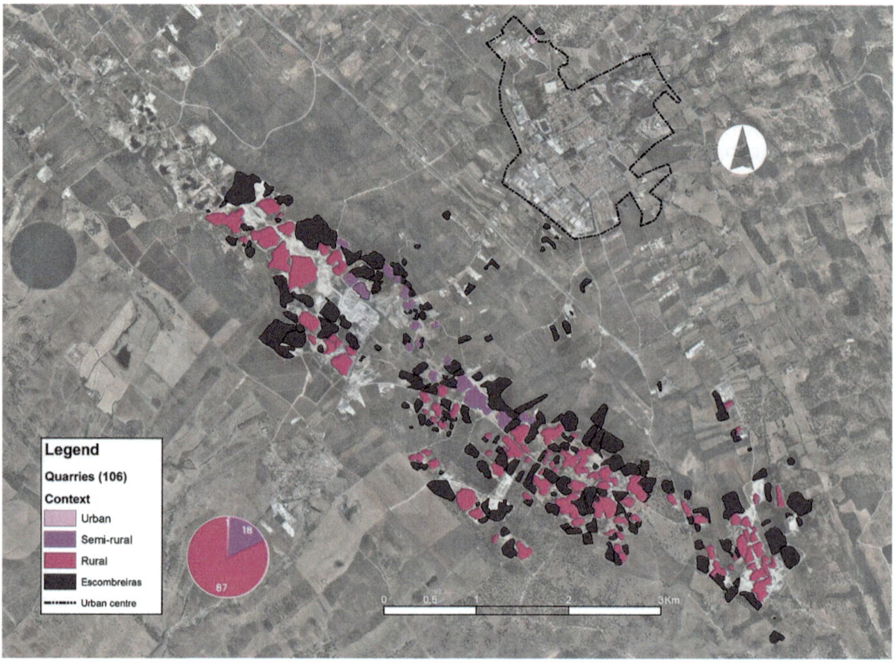

Fig. 5.27 Quarries' context. (Katia Talento, 2021)

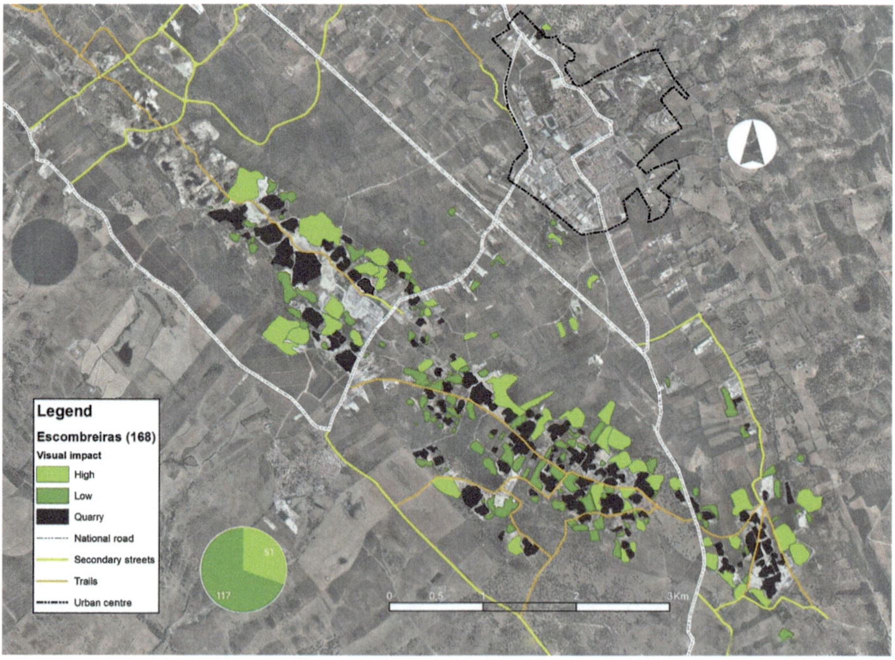

Fig. 5.28 *Escombreiras*' visual impact. (Katia Talento, 2021)

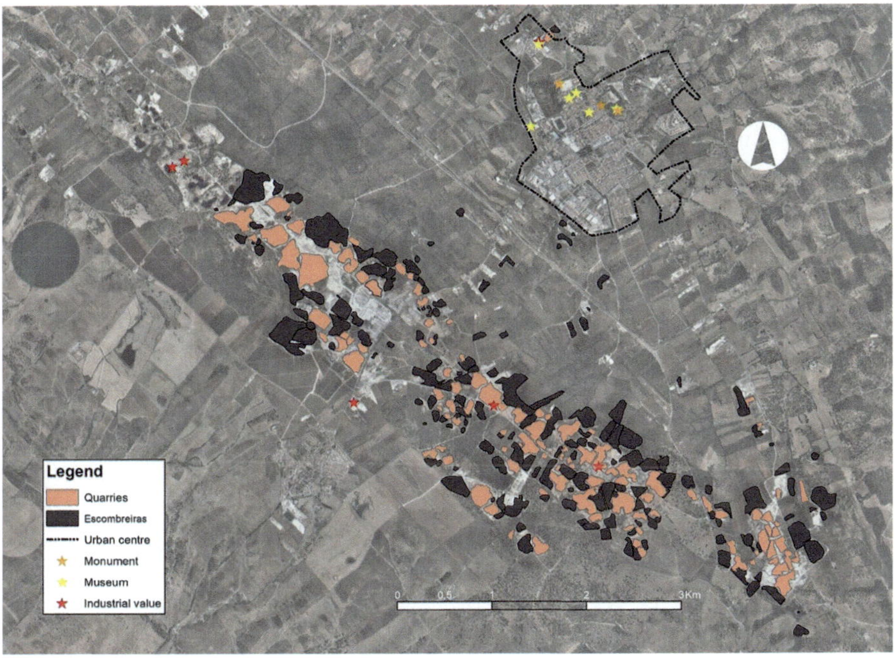

Fig. 5.29 Main touristic attractions of the municipality of Vila Viçosa. (Katia Talento, 2021)

The detailed analysis of the features of the quarries and *escombreiras* of Vila Viçosa, associated to the typology of rehabilitation and reuse to adopt in this municipality, has generated the planning of the industrial abandoned area. In this municipality, the location of the quarries and *escombreiras* respects the Directive Masterplan (PDM). Most of the extraction activity is concentrated far from the urban center, in the peripherical and rural surrounding. Only a small number of quarries is located in the urban context, in proximity to the Marble Museum. For this reason, the regeneration of this area consists in the planning of a **cultural and educational zone**, respecting the pre-existence. As happens in the other municipalities of the Estremoz Anticline (before described), the zoning of Vila Viçosa, in addition to the cultural zone, includes:

I. **Industrial and interpretational zone**: It respects the great concentration of active quarries mixed with some abandoned. Being the industrial activity still ongoing, the redesign of this area only involves few and small interventions, such as securing and the introduction of paths, rest stops, and viewpoints, recognizing the pre-existing industrial value (Fig. 5.30).
II. **Natural and agricultural zone with terraced gardens**: It represents a "pause" inside the industrial and interpretation zone. The green areas have been identified in the middle of the industrial park where only the abandoned quarries and *escombreiras* exist (Fig. 5.30).

Fig. 5.30 Planning of the industrial area of the municipality of Vila Viçosa. (Katia Talento, 2021)

III. **Economic zone**: The idea of considering the waste heaps as main elements of the rehabilitation process of the industrial area will improve the touristic flow. For this purpose, the planning operation includes accommodation facilities inside the piles of remains with moderate dimensions. The associated quarries, with an operation of filling, will turn into leisure places, such as swimming pools and bathing areas (Fig. 5.30).

5.5 Selection of an Intervention Subarea and Topographic Survey

5.5.1 Intervention Priorities and Criteria for the Evaluation of Quarries' and Escombreiras' Risk Condition

The proposed planning of the Estremoz Anticline, jointly with the elaboration of a set of primary guidelines for the rehabilitation of quarries and the reuse of waste heaps, on the one hand, allows to define intervention priority and, on the other hand, to design the best way to rehabilitate each abandoned unit. All without discontinuing the industrial activity in progress and including the reuse of waste heaps in the local. The intervention priorities are strongly influenced by the risk condition to which many quarries and *escombreiras* are subjected. The evaluation of the risk state is crucial for the selection of subareas in which to subsequently apply the architectural and landscape proposal. Due to the absence of means, the current research considers the risk situation of the quarries and *escombreiras*, through the analysis of the following features:

- State of activity
- Dimensions
- Distance to main streets (>50 meters according to the law)
- Distance to secondary streets (>15 meters according to the law)
- Visual impact
- Risk evaluated by the Portuguese General Directorate of Energy and Geology (DGEG)

This accurate evaluation has allowed to identify, in detail, which quarries and *escombreiras* are subjected to a risk condition for which it is necessary to take urgent action. Figure 5.31 shows the risk condition to which are exposed the quarries and waste heaps located in the municipality of Vila Viçosa. For a fluent reading of the study, the risk condition associated to the quarries and waste heaps of Estremoz and Borba municipalities is provided in the Annex of this book.

Based on this premise, the architectural and landscape proposal will be applied for a specific quarry and *escombreira*, belonging to the municipality of Vila Viçosa, with elevate risk (Fig. 5.31). For this purpose, as described in the following section, a meticulous topographic survey has been set up for the characterization of the selected subarea.

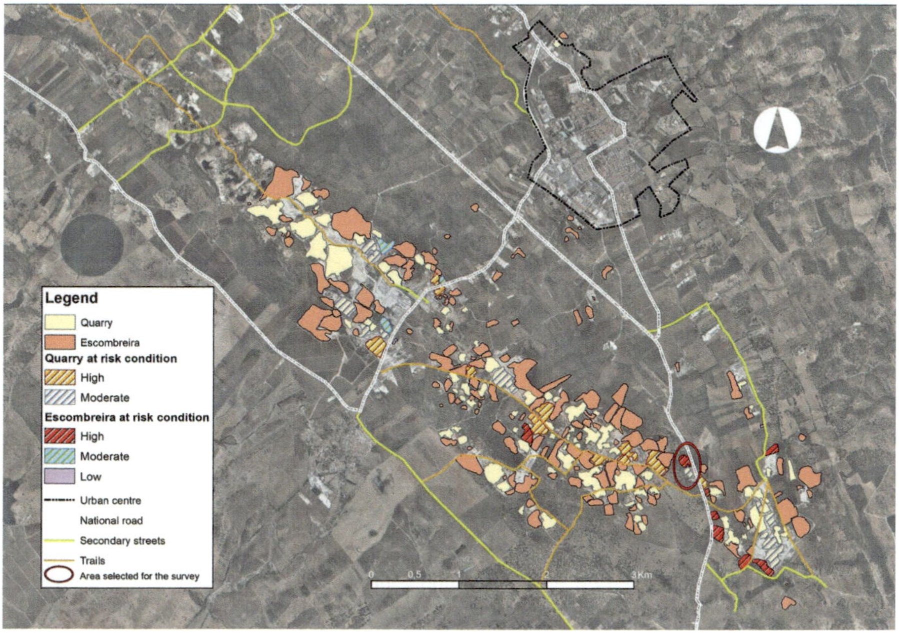

Fig. 5.31 Quarries and *escombreiras*, in Vila Viçosa, at risk condition. (Katia Talento, 2021)

5.5.2 Characteristics and Methodology of Field Data Acquisition for 3D Modeling

J. C. Kullberg
NOVA, Faculty of Sciences and Technology, Earth Sciences Department and GeoBiotec, University of Lisbon, Caparica, Portugal
e-mail: jck@fct.unl.pt

The topographic survey, carried out for the characterization of a specific subarea, located in the municipality of Vila Viçosa, has been extremely useful to obtain field data and then, to produce, from a photographic survey, a detailed and accurate 3D model of the selected quarry and the correspondent waste heap. The 3D model has been subsequently processed in Agisoft Metashape Software for the obtention of topographical plans and an orthophoto map, useful for the design of the architectural and landscape proposal. The survey has been set up with the use of the following equipment (Fig. 5.32):

- A DJI Phantom IV Pro drone with a 12.4 Mpx RGB camera (Fig. 5.32a) for the acquisition of georeferenced photographs of the ground.
- A Spectra GNSS SP60GIS high precision GPS (Fig. 5.32b) with a real time precision of 8 mm + 1 ppm in XY coordinates and 15 mm + 1 ppm in Z measures.

The main technical characteristics of the camera are as follows:

- Sensor—1/2.3" CMOS
- Effective pixels—12.4 M
- Lens—FOV 94° 20 mm (35 mm format equivalent) f/2.8 focus at ∞
- ISO Range—100–1600

Fig. 5.32 (**a**) Picture of DJI Phantom 4 Prof, an unmanned aerial vehicle (drone) used in the field to remotely acquire georeferenced vertical or oblique photographs. (**b**) Picture of the head of the Precision GPS device which has measured the position of ground control points, placed inside the area to fly. The area has been identified in several photographs taken from the drone to create a reliable 3D model. (Josè Carlos Kullberg and Katia Talento, 2021)

- Electronic shutter speed—8 to 1/8000 s
- Image size—4000 × 3000
- Still photography modes
- Single shot
- Burst shooting: 3/5/7 frames
- Auto-exposure bracketing (AEB): 3/5 bracketed frames at 0.7
- EV bias
- Timelapse
- HDR

The field data acquisition implies the use of ground control points (GCPs) (Fig. 5.33a–b) for a precise geopositioning to guaranty an accurate informatics processing to obtain 3D reliable modeling outputs. Thus, previously to the flight by the drone, 6 GCPs georeferenced by the SP60GIS have been distributed inside the area planned to be surveyed.

The aerial survey has been planned in the field to avoid eventual obstacles (such as aerial cables for energy transport), using the DJI's Ground Station Pro (GSP) software which communicates by wireless with the drone, and it controls the flights. In a wide variety of commands that controls both the flight and the quality of the photographs, which they will be processed in laboratory to obtain realistic, usable, and reliable models, it is of utmost importance to make the correct choice of the parameters of the flight and the range of values that must be used, depending on a

Fig. 5.33 (**a**) Positioning the GCPs in the field, to be pictured in the subsequent aerial photos taken by the drone. (**b**) Example of identification of one of the GCPs spread in the area flown by the drone. (Josè Carlos Kullberg and Katia Talento, 2021)

variety of intrinsic characteristics of the terrain and the environment. These characteristics can be shortened as follows:

- The topographic irregularity and the contrast and sharpness of the slopes, if they exist.
- The shape of the area to represent. Will it be regular or irregular? In the first case, it is good practice to consider the main orientation.
- The total surface area to optimize the batteries and the speed of the drone.
- The quality and intensity of sunlight, meaning the possible interference of clouds and the hour in the day.

In Fig. 5.34 (A1-2 and B1-2), the screenshots from the GSP software installed in an iPod show all the characteristics of different parameters used in the flight and the statistics, which result from the choices made.

It must be noted that in rough topographic surfaces, as this case due both the vertical walls of the quarry and the steep walls of the waste heap, surveys made with just one flight, with the camera in the vertical, are inappropriate because of the risk

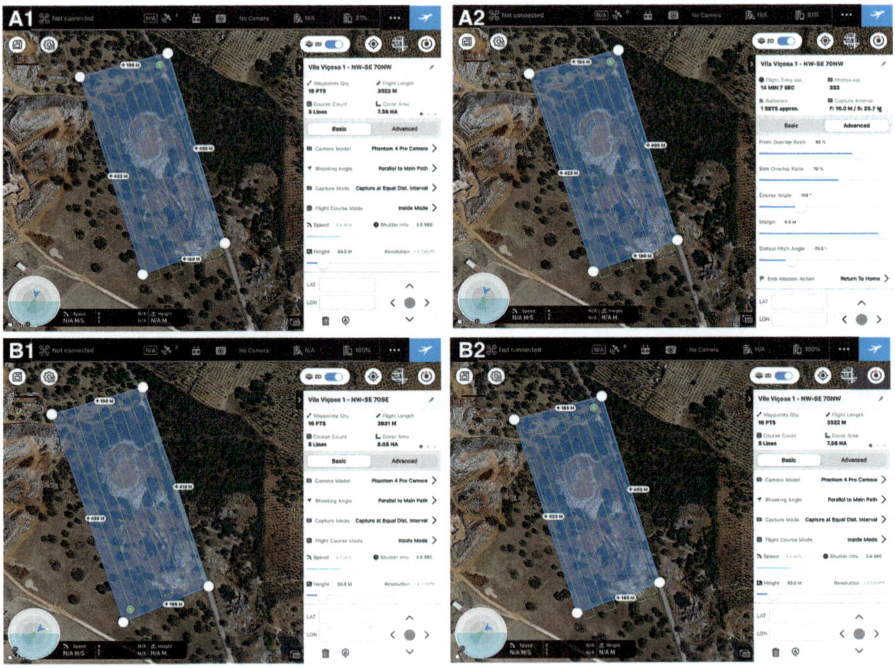

Fig. 5.34 (A-1, B-1) General view of the area and the flight plan over the quarry in the Vila Viçosa municipality; choose for the application of the architectural and landscape proposal. On the right side of images A2 and B2, the basic characteristics of the flight and statistics are illustrated. A and B correspond to two successive flights where the difference was the orientation of the pitch angle of the camera, meaning that for the A flight camera was deepening 70° to NW and for the B flight the camera was deepening 70° to SE (North to the top of the screenshots). (Josè Carlos Kullberg and Katia Talento, 2021)

of failing the representation of some walls is real. Thus, it is advised in these situations to make at least two flights with a tilted camera (70° used in this case) in two opposite directions. The green lines representing the path of the drone show exactly this effect used in the survey. In fact, as it can be seen in the A screenshots, the path is displaced outside to the SE of the area to be completely photographed, since the camera was deepening to NW. Otherwise, if the drone has flown strictly inside the blue area (they are to model), a strip parallel to the SE border of the area would not have been represented. The same applies, in the opposite direction, with the B flight. In what refers to the characteristics of the different parameters used in both flights, the following must be stressed (valid for flight A and B):

- Height of flight: 50 m above the place of take-off of the drone, approximately at the base level of the general original topographic surface. This implies a medium resolution of 1.4 cm/px for the photographs (see figs. A-1 and B-1).
- The speed of 5.0 m/s, a medium-low velocity was appropriate, at that height to prevent eventual drag and blurring of the photographs that would hinder a good quality of the informatics processing; it also allowed enough time (<2.0 s) for internal processing and saving in the external drive, a SD card, between two successive shots that were taken in intervals of 2.0 s ("Shutter Intv." in the "Basic" technical description).
- The camera was oriented parallel to the main path of the drone (the long green lines) and deepening (−) 70° (see in the "Advanced" characteristics). By convention, the "minus" signal (negative value) means that the camera was pointing downward, 0° being the horizontal position and the vertical pointing down = −90°.
- A fundamental choice must be made in the overlap ratios (both front and side overlaps) since they imply important "consequences" for the laboratory processing. On the one hand, if the overlaps are too high (>80%), the final number of photographs risks being extremely high to process, implying the need of extremely expensive computers for processing the orthophoto map and the 3D model; the generated point cloud can have billions of dots (EU unit). On the other hand, a low overlap (< 50%) allows very fast processing but, at the same time, it leads to models with low resolution and quality for 3D modeling. Since the area to model was not large, values between both "extremes" were choose. In the final data acquisition, 333 + 344 photos were obtained (see "Photos est." a statistic automatically calculated by the GSP software).

In general, the statistics calculated on the fly are:

- 16 Waypoints representing the number of times that the drone turned of direction, along the all path. Like the number of lines (see "Course Count"), they depend on the "side overlap ratio" and the height of the flight. The higher the flight, less lines are needed and the higher the overlap, more lines will be made for the complete path.
- Also depending on the previous parameters and values used, the flight length for A flight was 3522 m, and for B it was 3631 m. Apparently, both lengths should be equal; the difference consists in the different starting positions of both paths

("S" inside green circles) and the consequent differences in distance from the take-off position. This is a parameter that the user does not control, but it has some consequences on the time of flight, and consequently, the consumption of drone battery. In fact, for the longer length (B flight), the calculated time was 14 min 33 s, and for the shorter one (A flight), the time was 14 min 7 s.
- Since the image collection starts when the drone takes-off, the same effect in the statistics is confirmed.
- The total covered area (the blue polygon) was 8.05 ha (hectares).

As one can see in the following images, the result of the drone survey is represented by an orthophoto map (Fig. 5.35) and a topographic plan with level curves (Fig. 5.36) (derived from the image processing of an accurate 3D model of the area) of the selected subarea, composed of a quarry and the associated waste heap (Fig. 5.37).

5.6 Architectural and Landscape Rehabilitation Proposal

In the opening chapter, the introduction, the present research approached the topic of the rehabilitation of the Estremoz Anticline formulating three main questions. They can be resumed as follows:

Planning Main Question:

I. *Will it be possible to demonstrate the effectiveness of transforming a disused industrial landscape into an attractive, sustainable, and visited place through an operation of replanning?*

Landscape Main Question:

I. *Will it be achievable, through the elaboration of innovative sustainable strategies, to define the best form of rehabilitation of each landscape unity of the Estremoz Anticline?*

Architectural Main Question:

I. *Will it be feasible to involve the waste heaps in the process of rehabilitation of the area, creating original building and public spaces from them?*

The operation of replanning of the abandoned industrial area of the Estremoz Anticline, previously illustrated, can largely validate the first and the second question (planning and landscape questions), posed by this research work. In this regard, it has been scientifically demonstrated that, through the analyses of physical features of each quarry and *escombreira* and the elaboration of several typologies of reuse and rehabilitation, it is possible to create a formula of practical use to allocate new destinations of use for every abandoned territorial unit of the Estremoz Anticline. In this way, the original abandoned landscape can turn into a regenerated

Fig. 5.35 Orthophoto map of the surveyed subarea processed with Agisoft Metashape Software. (Katia Talento, 2021)

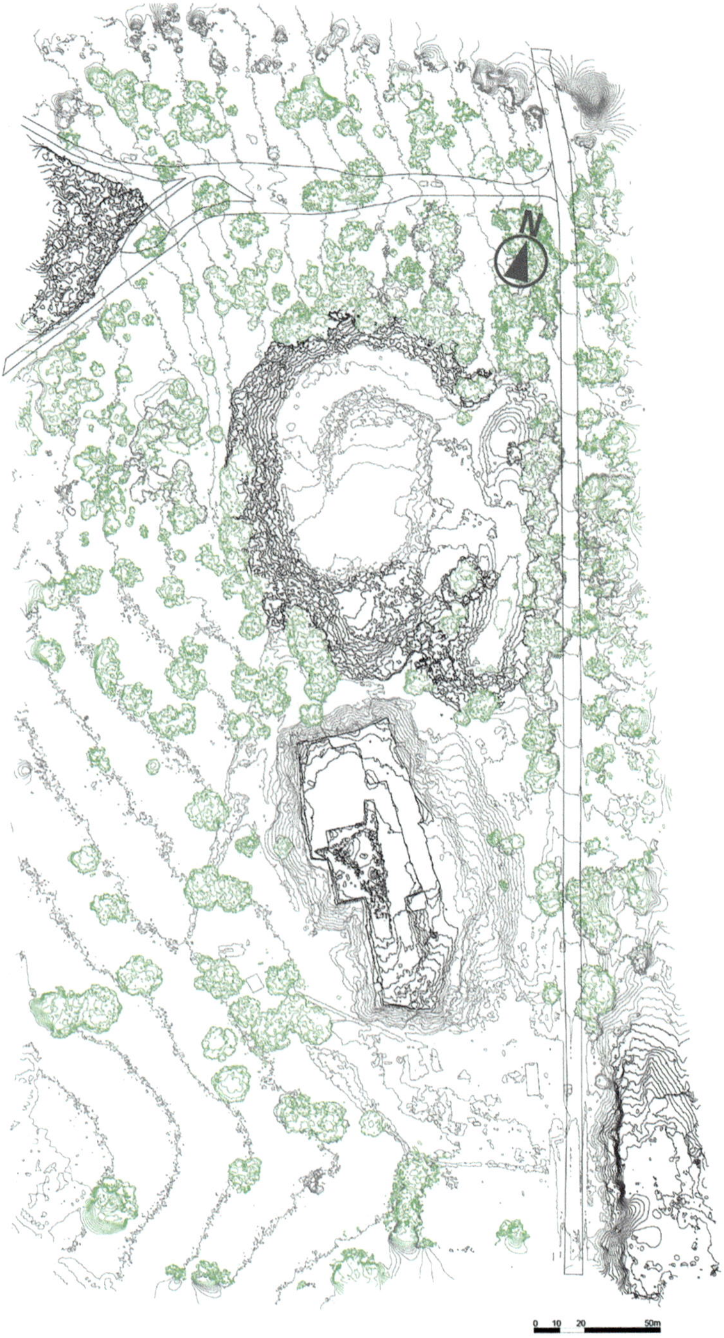

Fig. 5.36 Topographic map, with levels curves, of the surveyed subarea processed with Agisoft Metashape and Autocad Software. (Katia Talento, 2021).

Fig. 5.37 General view of the surveyed subarea composed of a quarry and the associated waste heap. (Katia Talento, 2021)

place, with renovated functions, having cultural, recreational, sport, economic, productive, touristic, environmental, and naturalistic purposes.

In the context of the present section, the architectural question posed by the research will be discussed and validated through the elaboration of an architectural and landscape proposal of rehabilitation of a specific subarea located in the municipality of Vila Viçosa. In accordance with the sustainable principles promoted by the entire research, which encourage the reuse of existing and residual materials to avoid the waste of available resources on Earth (increasingly scarce), the architectural and landscape proposal transforms the residual matter into a precious resource. For this reason, the waste material does not end its life cycle, but it is subjected to a "rebirth" becoming, jointly with the associated quarry, the principal actor of the process of rehabilitation.

This philosophy heals the wound originally inflicted on the landscape by creating a new identity and, at the same time, by restoring the existing identity, the industrial one. The philosophy of reuse represents a very powerful response to the economic values and identity crisis of the contemporary society, and it is a way of saying "enough." "Enough" with the accumulations, with wastes, with the exploitation of man by the environment, "enough" with the destruction of ecosystems, with this model of development which is not an evolution but an involution of society. The possibility to regenerate the quarries and reuse the waste heaps of the Estremoz

Fig. 5.38 Identification of the subarea selected for the architectural and landscape proposal in the municipality of Vila Viçosa. (Katia Talento, 2021)

Anticline, as largely discussed in the previous chapters, allows to achieve a circular economy, and it is strongly enhanced when this operation of reuse is carried out in the local, reducing the related transportation costs of the material and improving sustainability.

The subarea selected for the application of the architectural and landscape proposal belongs to the naturalistic zone with terraced gardens of the regeneration plan, proposed by this research work (Fig. 5.38).

For this area, the naturalistic typology of reuse is considered and, consequently, the waste heaps are treated as integral parts of the landscape and not as extraneous elements to contemplate. As previously resumed in Sect. 5.3, this typology of reuse allows the introduction of structures only if they appear "hidden" in the waste accumulation, generating a mimesis with the landscape. For the selected subarea located in the municipality of Vila Viçosa, the mimesis is obtained through the insertion of the building inside the waste heap, splitting it at the middle with a material "cut" (Fig. 5.39).

The "cut" is represented both in material and in metaphorical senses. In the context of the architectural and landscape proposal, the waste heap has been drastically "cut" as well as the landscape has been "cut" by human activity.

What does the "cut" operation symbolize in the architectural proposal?

It represents openings to let excerpts of the original landscape enter inside the *escombreira* to re-establish balances between the industrial and the natural side.

Fig. 5.39 Planimetry of the environmental interpretation center with the "cut" of the waste heap. (Katia Talento, 2021)

They represent new balances and life cycles that follow one another, restoring beauty to a landscape that for a long time has been the reason of exclusion and abandonment. The process of "cut" is expressed through a unique language, a common thread that embraces the waste heap and the quarry. This thread is materially visible, and it offers the visitor coherence, a visual guide, a sensorial path that exalts the monumentality and the importance of the pre-existence to let perceive the original damage created by human activity.

How brutal can human action be toward nature?

Through the current proposal, man crosses, both in physical and metaphorical senses, the wounds himself inflicted on the landscape, following a linear path generated by the "cut" of the waste accumulation. This path is composed of memory, history, industry, successes and failures, and wrong actions against the landscape and waste. It is a difficult path represented as a narrow tunnel in the middle of the waste heap, in which man feels a sensation of "heaviness," burden, impotence, and extra gravity because of the huge number of wastes located around and above him. This same path, toward the end of the waste accumulation, starts to lighten, to open toward the quarry; toward the rebirth, new meanings, and light; and toward the landscape, embracing sustainability.

Along this lengthy path, considered as a public space, the architectural proposal designs sustainable buildings including waste materials, parts of industrial machines to monumentalize, green islands that invade the industrial environment, and viewpoints to meditate, observe, and interpret the surrounding landscape. The landscape meaning and the industrial memory are expressed in their fullness, from the waste heap to the quarry, which together form a great environmental interpretation center, with a research center, a space dedicated to the promotion of the environmental and industrial heritage, a kind of ecomuseum that carries out an environmental awareness operation (Fig. 5.39).

The poetics of reuse covers all the angles of the proposal, not only through the reuse of the quarry and waste heap but also through the design of buildings and public space, reusing marble wastes for walls and pavements, second-hand materials (such as sheet metal) and industrial containers for facades, quarry belt for terraces and viewpoints, and unused industrial machines for the installations located along the public path inside the waste heap (Figs. 5.40, 5.41, 5.42, 5.43, 5.44, 5.45, 5.46, 5.47 and 5.48).

This proposal allows to improve sustainability because it provides for the reuse, in situ, of the local waste material, preserving the identity of the industrial place and obtaining energy savings because the wastes would not need to be moved elsewhere. In this way, one could have a nearly-zero emissions of CO_2 solution in respect of the principles discussed along the present research.

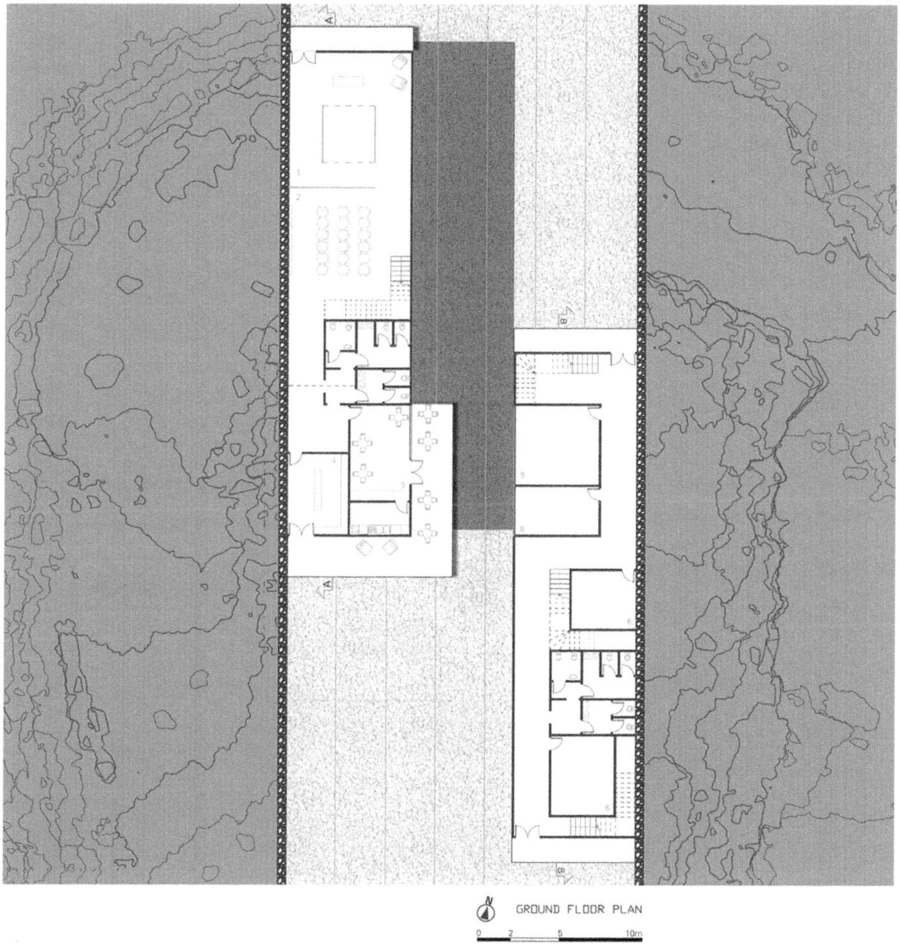

Fig. 5.40 Ground floor plan of the environmental interpretation center (Katia Talento, 2021) composed of: 1. Exhibition space, 2. audio-visual room, 3. café, 4. shop, 5. library, and 6. study room

Fig. 5.41 First floor plan of the environmental interpretation center (Katia Talento, 2021) composed of: 6. study room, 7. meeting room, 8. laboratory

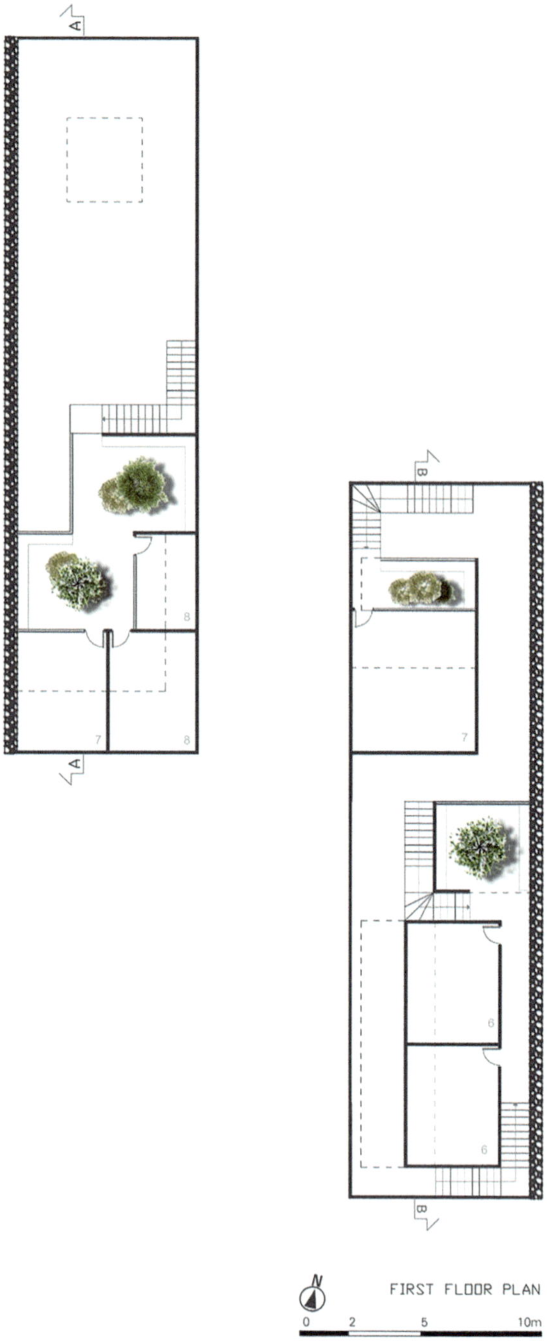

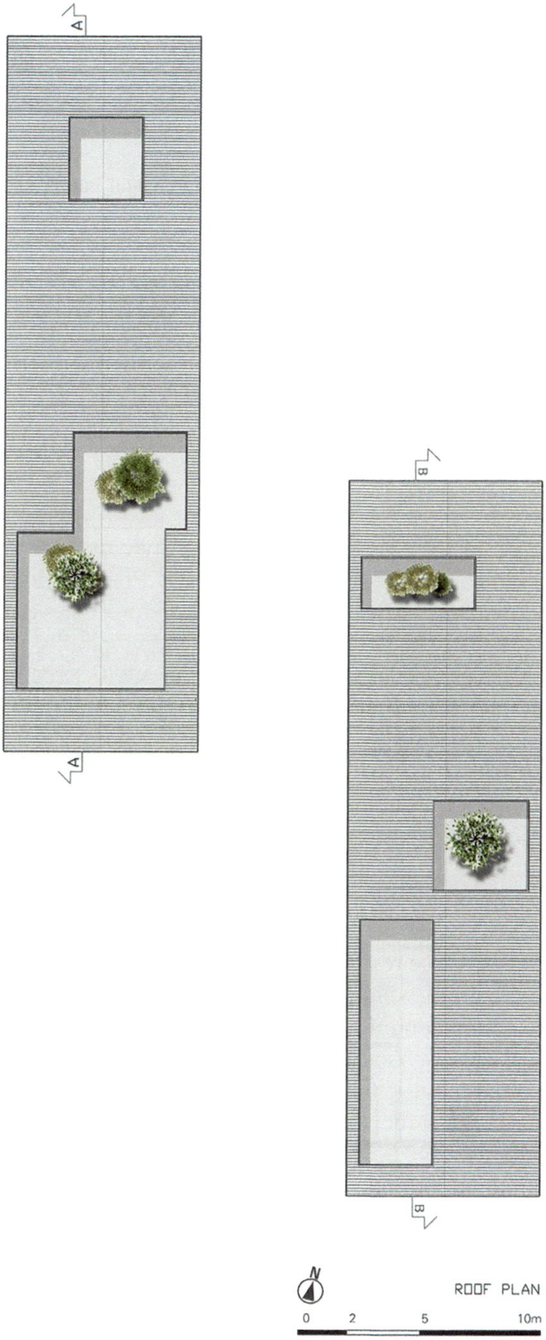

Fig. 5.42 Roof plan of the environmental interpretation center. (Katia Talento, 2021)

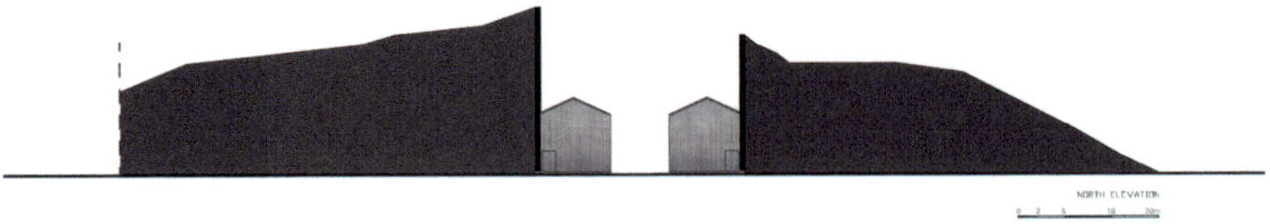

Fig. 5.43 North elevation of the environmental interpretation center inserted in the waste heap. (Katia Talento, 2021)

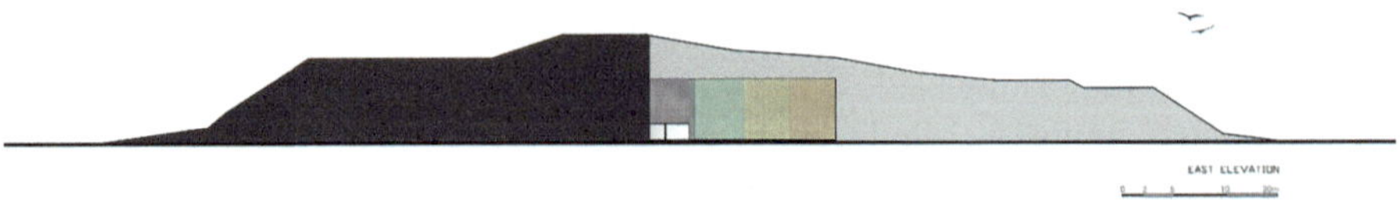

Fig. 5.44 East elevation of the environmental interpretation center, inserted in the waste heap. (Katia Talento, 2021)

Fig. 5.45 East elevation and Section B-B of the environmental interpretation center. (Katia Talento, 2021)

Fig. 5.46 West elevation and Section A-A of the environmental interpretation center. (Katia Talento, 2021)

Fig. 5.47 3D visualization of the environmental interpretation center exalting the "cut" of the waste heap and the connection with the associated quarry through an overhead path. (Katia Talento, 2021)

Fig. 5.48 3D visualization of the public space and the buildings created inside the waste heap. (Katia Talento, 2021)

5.7 Chapter Resume

The chapter that has just ended represents the core of the research, a chapter in which all the theoretical notions and the analysis of existing case studies, originally elaborated, converge in a unique practical outcome. After having presented the case study of the research and describing it in all its details, both physical and geological, the current chapter has shown the criteria according to which the quarries and the associated waste heaps, of the Estremoz Anticline, have been classified.

These criteria, jointly with the elaboration of original typologies of reuse, gave rise to a scientific formula of practical use, useful to confer new destinations of use to each abandoned quarry and waste heap. The formula has also set the basis for the elaboration of general guidelines, paving the way for all planners dealing with the problem of an abandoned extractive area needing to be regenerated.

As described in the chapter, the typologies of rehabilitation and reuse, the scientific formula, the criteria, and the general guidelines have allowed to create a complete replanning of the industrial area of the Estremoz Anticline, transforming it a different place, a sustainable place where tourism and culture are promoted, agricultural production is valorized, sports is encouraged, and natural landscape is completely revisited.

After having described the methodological process of the survey, made with the auxilia of a drone, for the selection of a specific quarry and waste heap of the Estremoz Anticline, in this chapter, an architectural and landscape proposal has been designed to demonstrate the possibility of including an inert accumulation of wastes in the process of regeneration of a quarry, turning it into a component perfectly integrated with the landscape and capable to acquire new meanings.

Bibliography

Alves, D. (2015). *Mármore, património para o Alentejo: Contributos para a sua história (1850–1986).* Centro de Estudos de Cultura, História, Artes e Patrimónios.

Araújo, A., Piçarra, J., Borrego, J., Pedro, J., & Oliveira, J. T. (2013). As regiões central e sul da Zona de Ossa-Morena. In R. Dias, A. Araújo, P. Terrinha, & J. C. Kullberg (Eds.), *Geologia de Portugal* (Vol. 1, pp. 509–549). Escolar Editora.

Baganha, P. (2015). *Notes towards a project for the quarries landscape of Estremoz anticline. Joeleho* (p. 6). Revista de Cultura Arquitectónica.

Cohen, D. (2007, May). *Earth's natural wealth: An audit.* https://www.newscientist.com/article/mg19426051-200-earths-natural-wealth-an-audit/?ignored=irrelevant

Cohen, K. M., Finney, S. C., Gibbard, P. L., & Fan, J. X. (2013). The ICS international chronostratigraphic. *Chart, 36*, 199–204.

Costa, C. (1992). *As pedreiras do Anticlinal de Estremoz. A Geologia de Engenharia na Exploração e Recuperação Ambiental de Pedreiras.* [PhD Thesis]. FCT/UNL.

De Poli, M., & Incerti, G. (2014). *Atlante dei paesaggi riciclati.* Skira editeore.

DGEG – Divisão de Pedreiras do Sul. (2016). Caraterização do setor da extração de massas minerais. *Situação face à regulamentação técnica e legal.*

Geoportal. (n.d.). *Geoportal.* Retrieved 26 July 2022, from https://geoportal.lneg.pt/pt/bds/rop/

Gonçalves, F. (1970). Contribuição para o conhecimento geológico dos mármores de Estremoz. *Notas e Trabalhos Do SFM, 20*(1–2), 201–209.

Gonçalves, F. (1974). Notícia explicativa da folha 37-C (Juromenha). In *Geological map of Portugal in the 1/50 000 scale.* Serviços Geológicos de Portugal.

Gonçalves, F., & Carvalhosa, A. (1994). O Proterozóico da Zona de Ossa Morena no Alentejo. Síntese e actualização de conhecimentos. *Mem.A Cad. Ciênc, 34*, 141–173.

Gutiérrez-Marco, J. C., Piçarra, J. M., Meireles, C. A., Cózar, P., García-Bellido, D. C., & Pereira, Z. (2019). Early Ordovician–Devonian passive margin stage in the Gondwanan units of the Iberian Massif. In *The geology of Iberia: A geodynamic approach* (Vol. 2, pp. 75–98). Springer.

Ladeira, F. L. (1977). Mármores. *Boletim de Mina, 14*(4), 195–199.

LNEG. (2020). *Mármores e calcários ornamentais de Portugal.* Atelier Nunes & Pã.

Lopes, J. L. G. (2003). Contribuição para o conhecimento Tectono–Estratigráfico do Nordeste Alentejano, transversal Terena—Elvas. Implicações económicas no aproveitamento de rochas ornamentais existentes na região (Mármores e Granitos).

Lopes, L. (2020). Anticlinal de Estremoz: Geologia, Ordenamento do Território e Produção de Rochas Ornamentais após 2000 Anos de Exploração. *Boletim de Minas, 54*, 59–83.

Lopes, M., & Martins, R. (2015). Global heritage stone: Estremoz Marbles, Portugal. In *Global heritage stone: Towards international recognition of building and ornamental stones* (Vol. 407, pp. 57–74). Geological Society Sp.

Manuppella, G., Moreira, J. B., & Romão, M. L. (1981). Calcários Portugueses Sua Utilização Industrial. *Boletim de Mina, 18*(4), 271–283.

Moreira, J., & Vintém, C. (1997). *Geological Map of the Estremoz Anticline in the 1/25.000 scale.* Instituto Geológico e Mineiro.

Moura, A. C., & Martins, O. R. (1992). *Catálogo de Rochas Ornamentais Portuguesas (Vols 1–4).* Min. Ind. Energia.

Oliveira, V. (1984). Contribuição para o conhecimento geológico-mineiro da região de Alandroal-Juromenha (Alto Alentejo). *Estudos, Notas e Trabalhos Do SFM, 26*(1–4), 103–126.
Perdigão, J. C. (1976). *Notícia explicativa da folha 37-C (Juromenha), geological map of Portugal in the 1/50 000 scale*. Serviços Geológicos de Portugal.
Pereira, M. F., Solá, A. R., Chichorro, M., Lopes, C. L., Gerdes, A., & Silva, J. B. (2012). North-Gondwana assembly, break-up and paleogeography: U–Pb isotope evidence from detrital and igneous zircons of Ediacaran and Cambrian rocks of SW Iberia. *Gondwana Research, 22*(3–4), 866–881. https://doi.org/10.1016/j.gr.2012.02.010
Piçarra, J. M., & Le Meen, J. (1994). Ocorrência de crinóides em mármores do Complexo Vulcano-Sedimentar Carbonatado de Estremoz: Implicações estratigráficas. *Instituto Geológico e Mineiro, 80*, 15–25.
Piçarra, J. M., & Sarmiento, G. (2006). *Problemas de posicionamento estratigráfico dos Calcários Paleozóicos da Zona de Ossa Morena (Portugal)* (pp. 657–660).
Piçarra, J. M., Sarmiento, G. N., & Gutiérrez-Marco, J. C. (2014). Geochronological vs. Paleontological dating of the Estremoz marbles (Ossa Morena zone, Portugal)—New data and reappraisal. In *Gondwana 15 north meets south* (Vol. 140). Instituto Geológico y Minero de España.
Quesada, C., & Oliveira, J. T. (2019). *The geology of Iberia: A geodynamic approach* (Vol. 2). Springer.
Ribeiro, T., Vendas, D., Caetano, P. S., Costa, C. N., & Rocha, M. (2010). Metodology for characterization of marble quarry tailings piles from the Marble Zone (Alentejo) for evaluation of reprocessing potencial. *E-Terra, 20*(18).
Shirazi, E. K. (2011). Reusing of stone waste in various industrial activities. 4.
Talento, K., Amado, M., & Kullberg, J. C. (2019). Landscape—A review with a European perspective. *Land, 8*(85). https://doi.org/10.3390/land8060085
Ulubeyli, G. C., & Artir, R. (2015). Properties of hardened concrete produced by waste marble. *Procedia Social and Behavioral Sciences, 195*, 2181–2190.
Use it more. Superuse Studios. (n.d.). Arch20. https://www.arch2o.com/use-superuse-studios/
Wong, M. H. (1986). Reclamation of wastes contaminated by copper, lead, and zinc. *Environmental Management, 10*, 707–713. https://doi.org/10.1007/BF01867725

Part IV

Chapter 6
Conclusions and Future Developments

Reuse the past, recycle the present, save the future.

Anonymous

6.1 Main Book Conclusions

Over the years, human activity has drastically mutated the ecosystem of our planet, and many changes have been flanked by life-threatening hazards. The complexity of human needs for natural resources has considerably increased with the growing levels of population, production, industrialization, and pollution causing an uncontrolled and exaggerated use of resources and, subsequently, the "greenhouse effect". The current research has provided a fundamental contribution to promote sustainable development with the elaboration of new architectural strategies in terms of sustainability, focused on the reuse process of existing and discarded materials.

The **first part** of the research, presented in this book, corresponds to the introduction in which one has been reflected on the sustainability concept and the damages caused by man on Earth. In this context, the architecture field plays a paramount role, that of compensating for these damages and the resources spent for the construction of buildings. Nowadays, architecture is strongly influenced by sustainability principles, and it aims to create healthy living environments decreasing the use of human resources and, at the same time, improving the use of renewable energies, natural materials, and the reduction of wastes, through the reuse and recycle of local discarded materials. Wastes have always been identified as undesired materials, but, through the current research, they have been considered innovative elements to redesign architectures and public spaces, redefining the meaning of an abandoned place. Considering this overview, the first part of the study has further defined the background, which has provided insight on quarries (main topic of the study), the research sphere, scope, problematics, objectives, questions, hypothesis, and methodology.

K. Talento, *Sustainable Buildings from Marble Waste Heaps*,
https://doi.org/10.1007/978-3-031-72146-5_6

The state of the art, developed in the **second part** of the research, represents a consistent and validated analysis of the main topics underlying the scope and question of the research: the landscape mutation and the rehabilitation of quarries. **Chapter 2**, **Landscape Mutation**, has provided succinct theories about landscape to demonstrate the versatility of the natural element, belonging to various fields of intervention. At the same time, the chapter has described a landscape strongly humanized, due to the intensive and uncontrolled industrialization, witnessing the birth of the concept of "Drosscape", made of industrial machines, wastes, and abandoned landscapes, after being widely exploited.

As part of the state of the art, **Chap. 3, The Rehabilitation of Quarries**, has focused on the extraction activity and the reconversion of industrial abandoned places to something with architectural and landscape value. As shown in the chapter, by means of several examples, quarries have been converted into large public parks, botanic gardens, natural areas of considerable environmental value, auditoriums, tourism buildings, infrastructures, shopping malls, and cemeteries. Based on these case studies, the chapter has elaborated a schematization containing existing reuse typologies of a quarry and their discussions, which successively add new methodologies of rehabilitation, never before experienced.

The **third part** of the research has addressed the notion of waste reuse, circular economy, and the problem associated with the massive accumulations of waste from extraction activity. **Chapter 4**, **Waste Reuse from Quarries**, after having provided different meanings of wastes and discussed the importance of the circular economy theory, has deeply analysed several case studies where the reuse of waste material from quarries is crucial, classifying them according to two distinct methodologies of reuse: off-site reuse and reuse in situ. Being the main topic of the research, the current chapter has paid attention to the second methodology, and it has sequentially culminated in the elaboration of innovative typologies of reuse, setting up the basis for the elaboration of a systematic methodology of reuse to apply to a practical case study in Portugal.

Finally, **Chap. 5**, **Case Study Application**, has presented the Portuguese case study, located in the Estremoz Anticline, and it proposed new sustainable strategies for the replanning of the area and the design of an architectural and landscape proposal, in accordance with the reuse principles. These strategies were possible thanks to the elaboration of an original formula that has conferred new destinations of use to the abandoned quarries and waste heaps of the Estremoz Anticline. This scientific formula has also allowed to create the criteria and a primary set of guidelines for planners, for the integrated reuse of an extractive area.

In terms of reaching the proposed objectives and challenges:

- The general objective of requalifying the extractive landscape of the Estremoz Anticline, with elements and materials available in situ, has been fully reached through the operation of replanning of the entire area.
- It has been clearly demonstrated, in Chap. 5, that it is possible to define the perfect form of regeneration of each landscape unity, conferring beauty, harmony, and functionality to the abandoned industrial landscape and proving that the

huge accumulations of waste will no longer be a burden, with negative impact, in the landscape.
- As shown in Chap. 5, it has been well succeeded the attempt to put into practice the concept of waste reuse, through the proposal of architectures and public spaces built inside the mounds of wastes, linking them to the corresponding quarries and the surrounding landscape.

Besides the enumerated reached goals, the possibility of having general guidelines for the rehabilitation of quarries and the reuse of waste heaps, in specific situations, has also been achieved. These guidelines are extremely important and useful for the regeneration of any abandoned extractive site, and they derive from the elaboration of a new scientific formula of practical use, as previously illustrated in Chap. 5. Although this part of the work has been partially published (in several journal articles), there was a significant improvement in the rehabilitation guidelines and in the elaboration of the scientific formula for the regeneration of quarries. Therefore, from the author's point of view, this part of the work represents one of the most important contributions of this research.

If one lesson should be taken from this book, it is that the mounds of wastes, located in the industrial abandoned areas, should be treated as elements with a great potential of reuse, in situ, and not as materials to be suppressed. Several places in the world are suffering from problems related to the huge accumulation of remains in the industrial context. The current investigation could represent some advice to deal with these problematics, present in different surroundings, through the application of original approaches.

6.2 Future Developments

In the continuation of the present study, further research can be carried out in various aspects of the rehabilitation of quarries and the reuse of waste heaps. In particular, the following aspects seem extremely promising to develop in the corresponding domains:

- **Geological**: to evaluate the risk of potential rock movements and instabilities using a structural and geological characterization of quarries and *escombreiras*
- **Engineering**: to study the best structural solution for the securing and the "cut" of the waste heap and the creation of a public space inside it, composed of low-cost buildings, wastes from the industrial activity, and natural space
- **Environmental**: to prove that the huge accumulations of wastes will no longer be a burden, with a negative impact, on the landscape, evaluating:
  - The new environmental impact
  - The study of an innovative methodology to avoid the loss of biodiversity, caused by the excavation processes, which modify the ecosystems

- **Social**: to validate the planning process and the architectural proposal, elaborated in the context of the present research, with local municipalities, regional development committee, and institutions and research centres that deal with the problems of waste reuse in landscape and architecture:
  - To re-establish contacts between the local population and the abandoned industrial area
  - To improve the social net by enhancing the tourism in the degraded territory
- **Economic**: to study cost-effective methodologies for the creation of architectures and public spaces built inside the mound of wastes:
  - To demonstrate that the operation of leaving the waste heaps will be more economically advantageous than the other one of demolishing them

Annex

Estremoz: Characterization of Quarries and Escombreiras

State of Activity (Figs. A1 and A2)

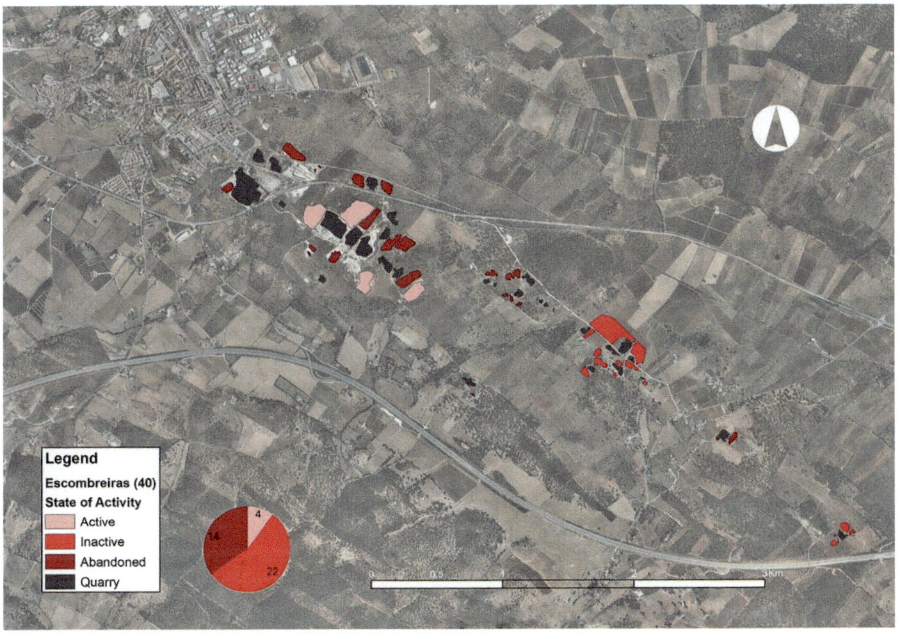

Fig. A1 *Escombreiras*' state of activity. (Katia Talento, 2021)

K. Talento, *Sustainable Buildings from Marble Waste Heaps*,
https://doi.org/10.1007/978-3-031-72146-5

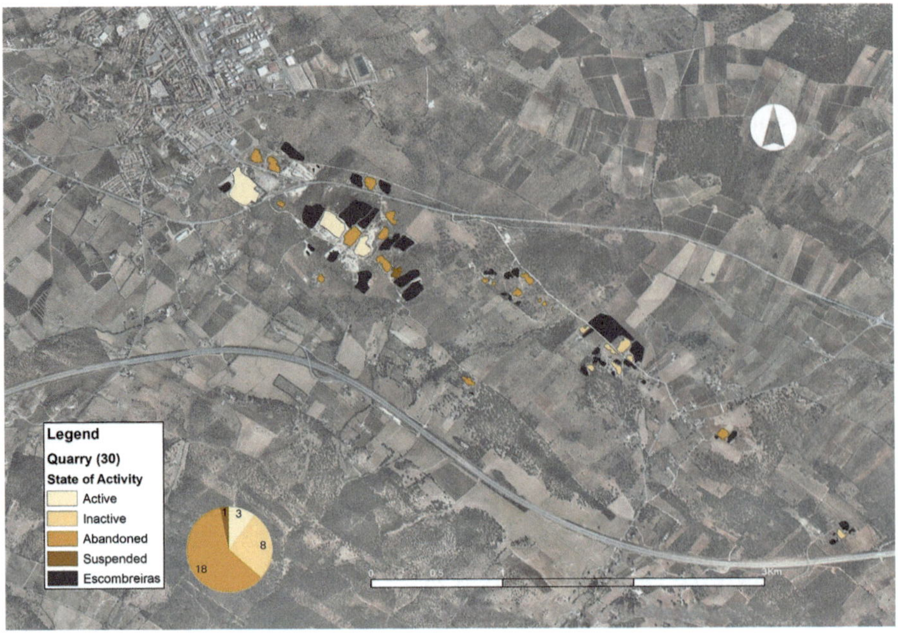

Fig. A2 Quarries' state of activity. (Katia Talento, 2021)

Dimensions (Figs. A3 and A4)

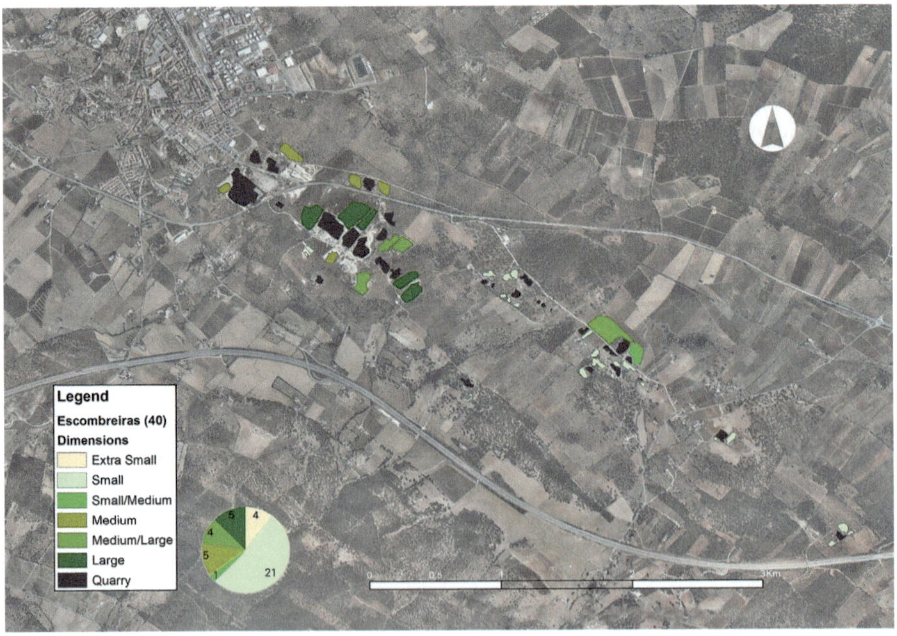

Fig. A3 *Escombreiras*' dimensions. (Katia Talento, 2021)

Fig. A4 Quarries' dimensions. (Katia Talento, 2021)

Proximity to Main and Secondary Streets (Figs. A5, A6, A7 and A8)

Fig. A5 *Escombreiras*' proximity to main streets. (Katia Talento, 2021)

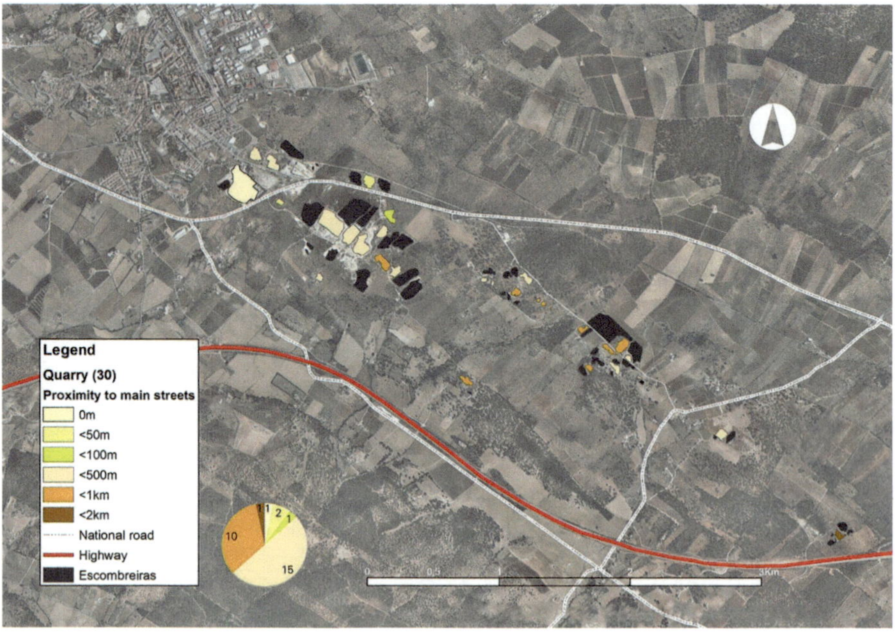

Fig. A6 Quarries' proximity to main streets. (Katia Talento, 2021)

Fig. A7 *Escombreiras'* proximity to secondary streets. (Katia Talento, 2021)

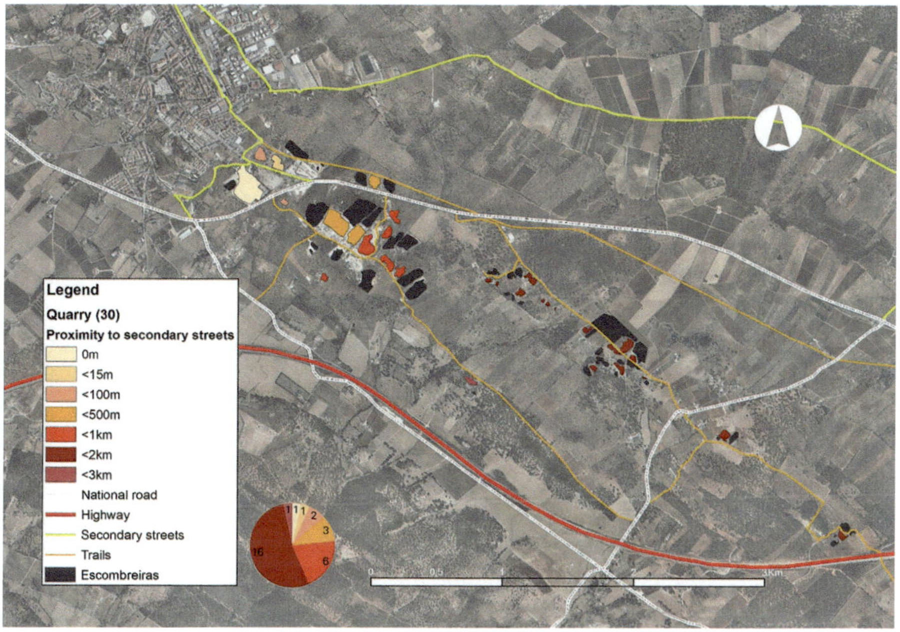

Fig. A8 Quarries' proximity to secondary streets. (Katia Talento, 2021)

Proximity to the Urban Centre (Figs. A9 and A10)

Fig. A9 *Escombreiras*' proximity to the urban centre. (Katia Talento, 2021)

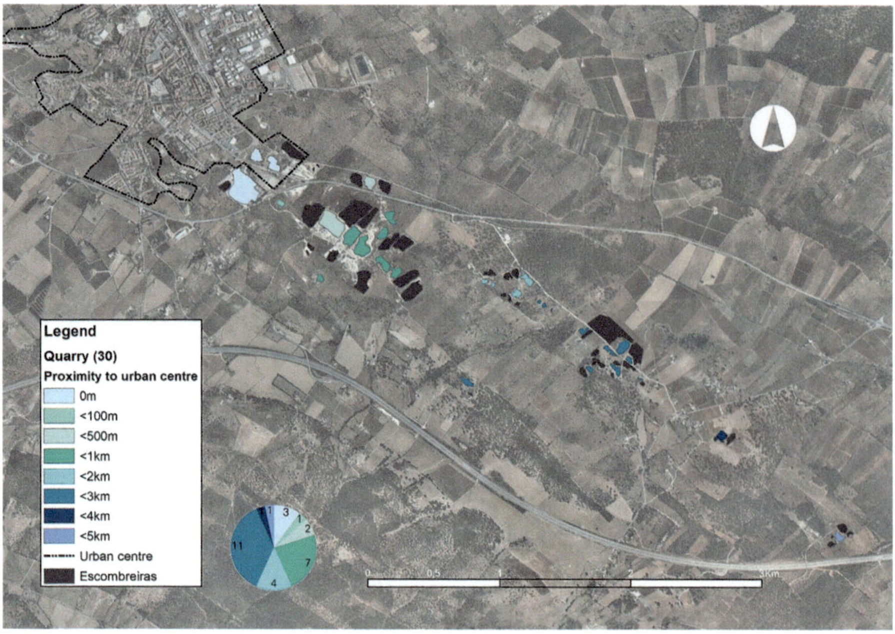

Fig. A10 Quarries' proximity to the urban centre. (Katia Talento, 2021)

Context (Figs. A11 and A12)

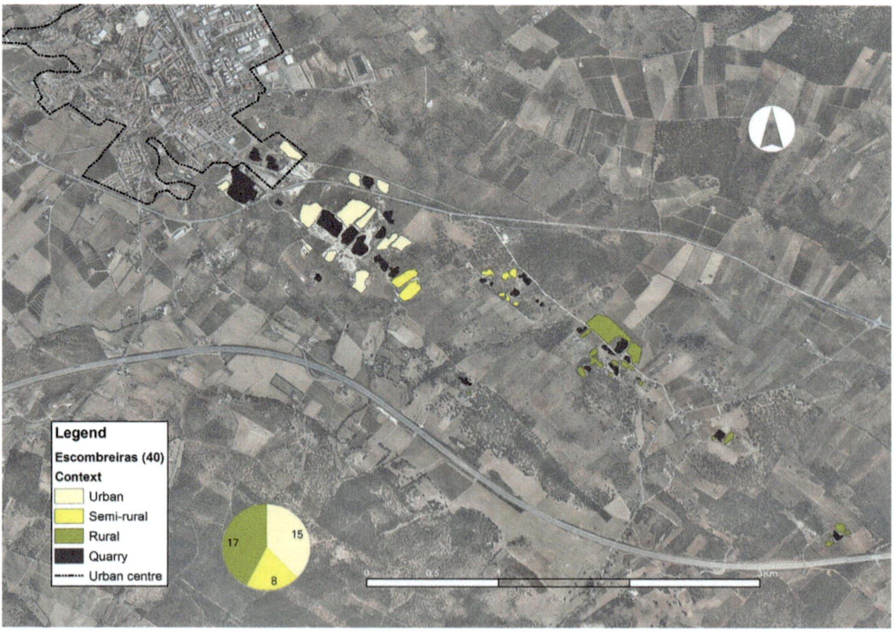

Fig. A11 *Escombreiras'* context. (Katia Talento, 2021)

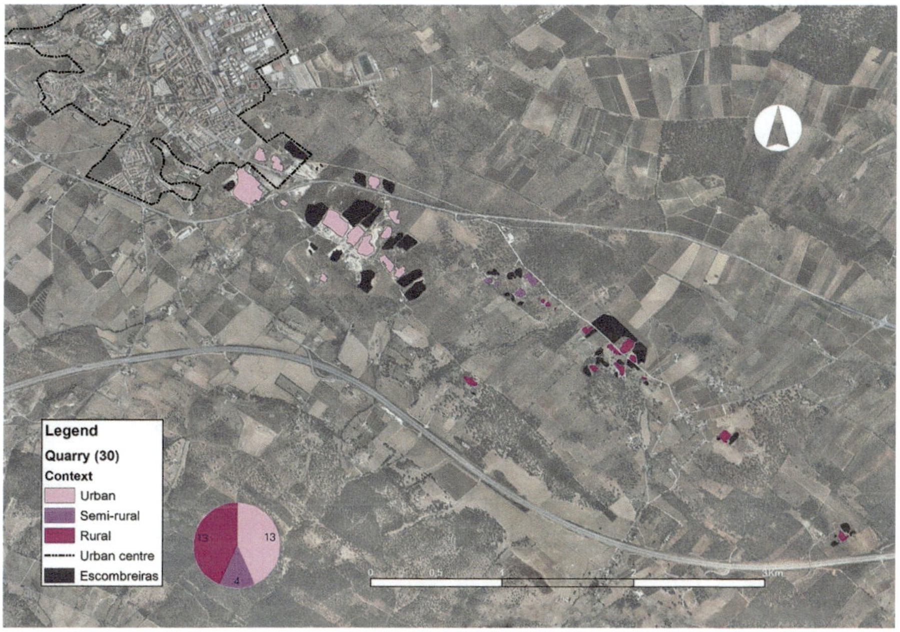

Fig. A12 Quarries' context. (Katia Talento, 2021)

Visual Impact (Fig. A13)

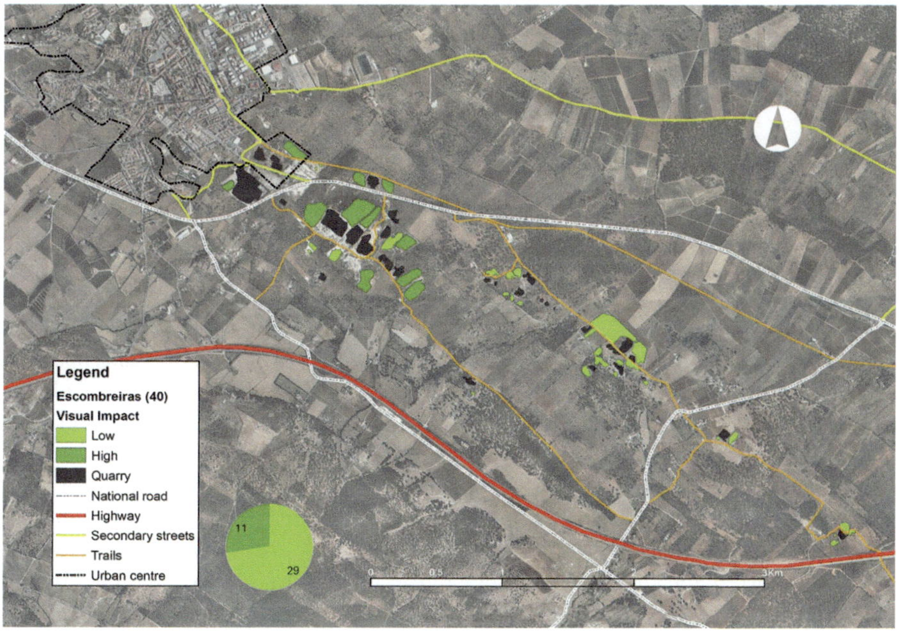

Fig. A13 *Escombreiras*' visual impact. (Katia Talento, 2021)

Main Attractions (Fig. A14)

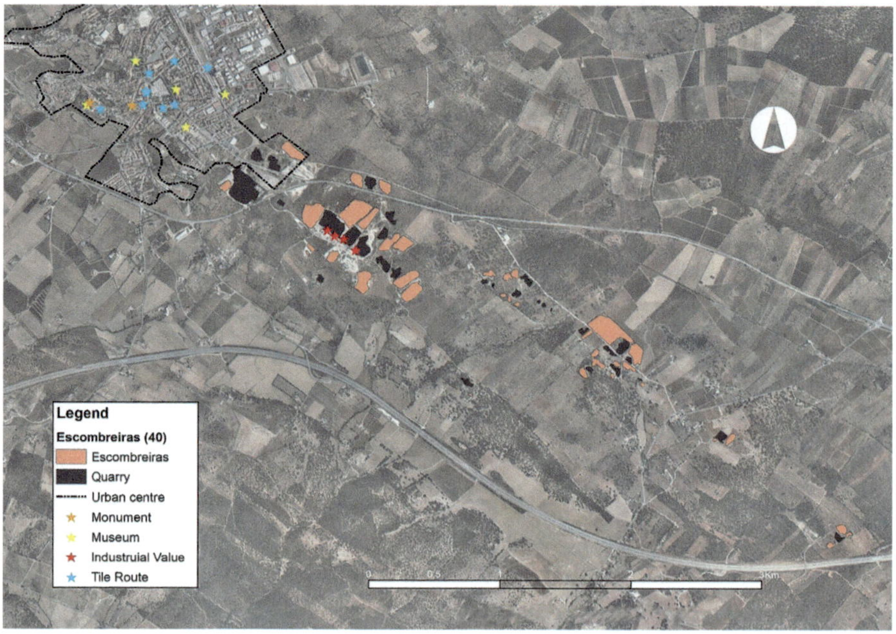

Fig. A14 Main attractions of Estremoz. (Katia Talento, 2021)

Risk Condition (Fig. A15)

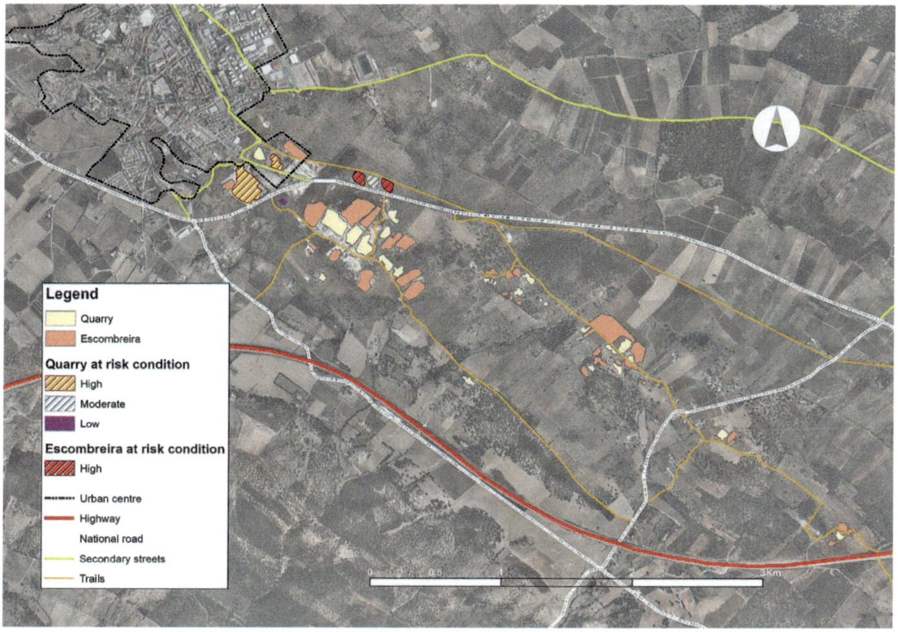

Fig. A15 *Escombreiras* and quarries subjected to a risk condition. (Katia Talento, 2021)

Borba: Characterization of Quarries and Escombreiras
State of Activity (Figs. A16 and A17)

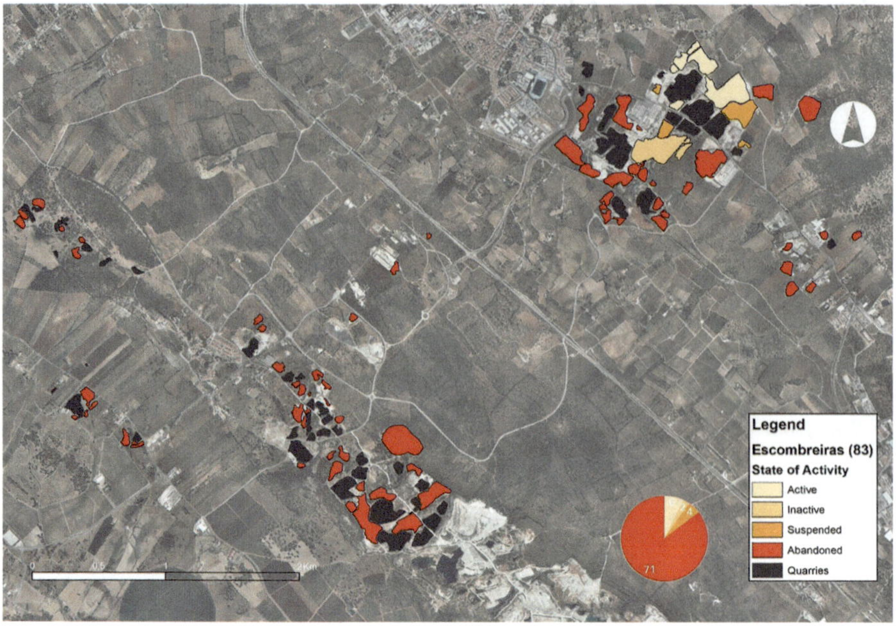

Fig. A16 *Escombreiras*' state of activity. (Katia Talento, 2021)

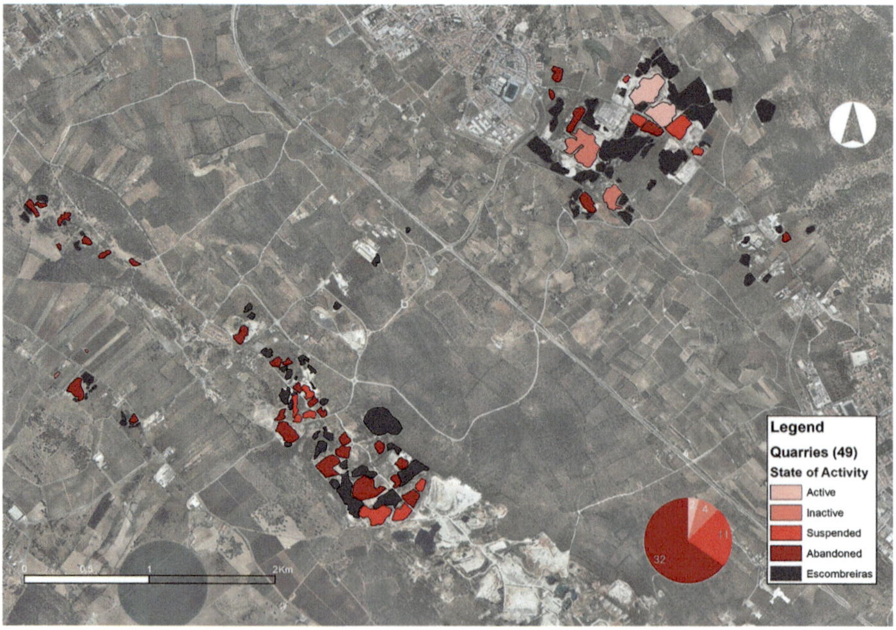

Fig. A17 Quarries' state of activity. (Katia Talento, 2021)

Dimensions (Figs. A18 and A19)

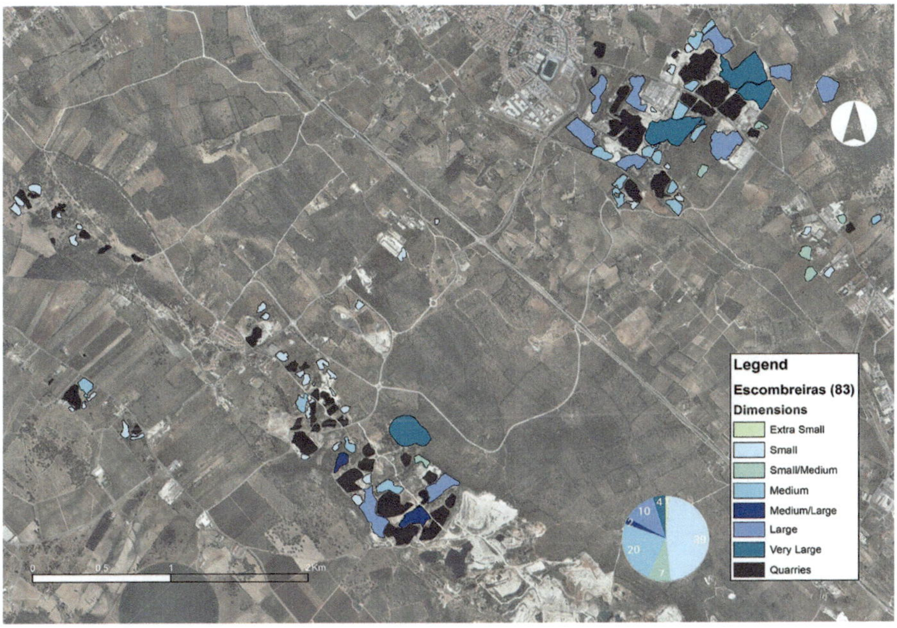

Fig. A18 *Escombreiras*' dimensions. (Katia Talento, 2021)

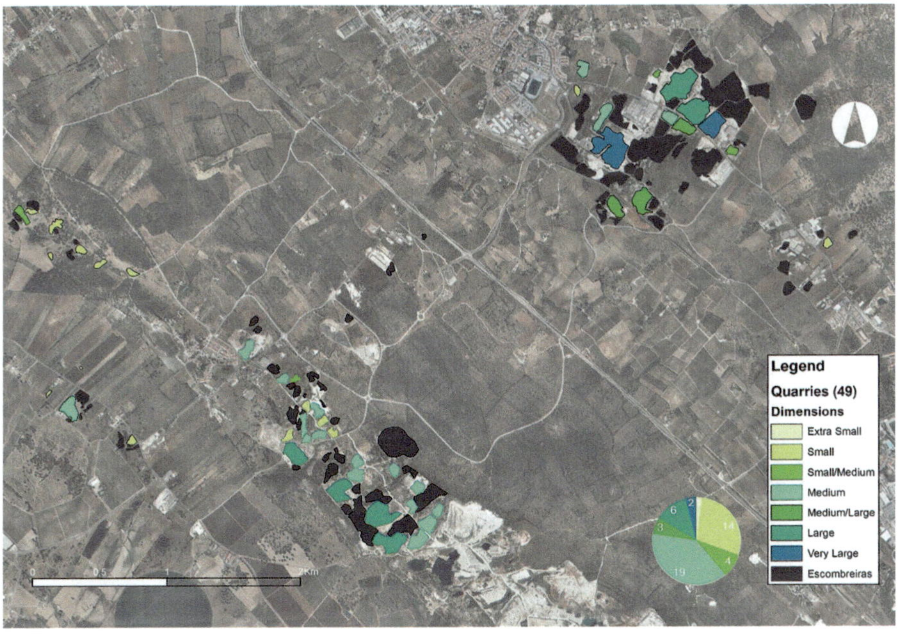

Fig. A19 Quarries' dimensions. (Katia Talento, 2021)

Proximity to Main and Secondary Streets (Figs. A20, A21, A22, and A23)

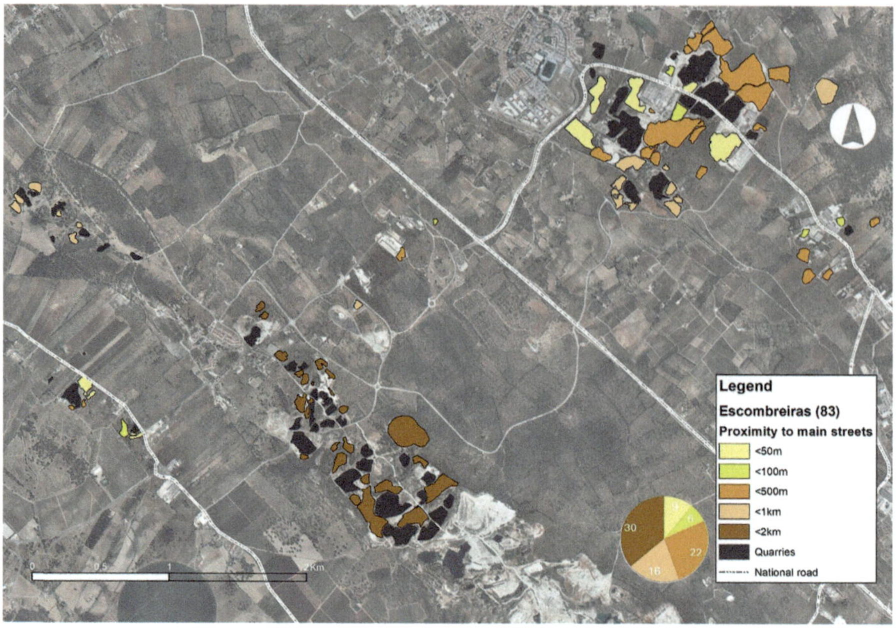

Fig. A20 *Escombreiras*' proximity to main streets. (Katia Talento, 2021)

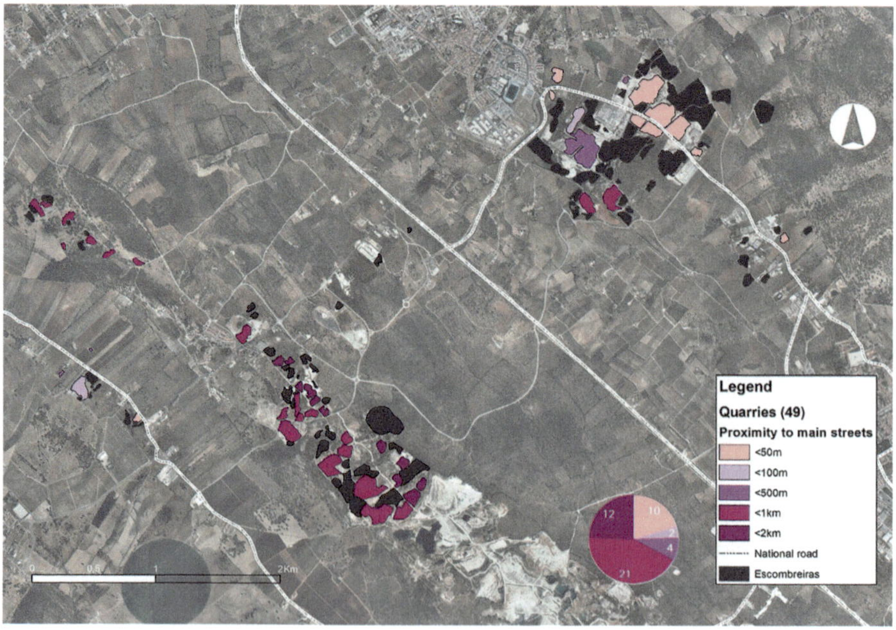

Fig. A21 Quarries' proximity to main streets. (Katia Talento, 2021)

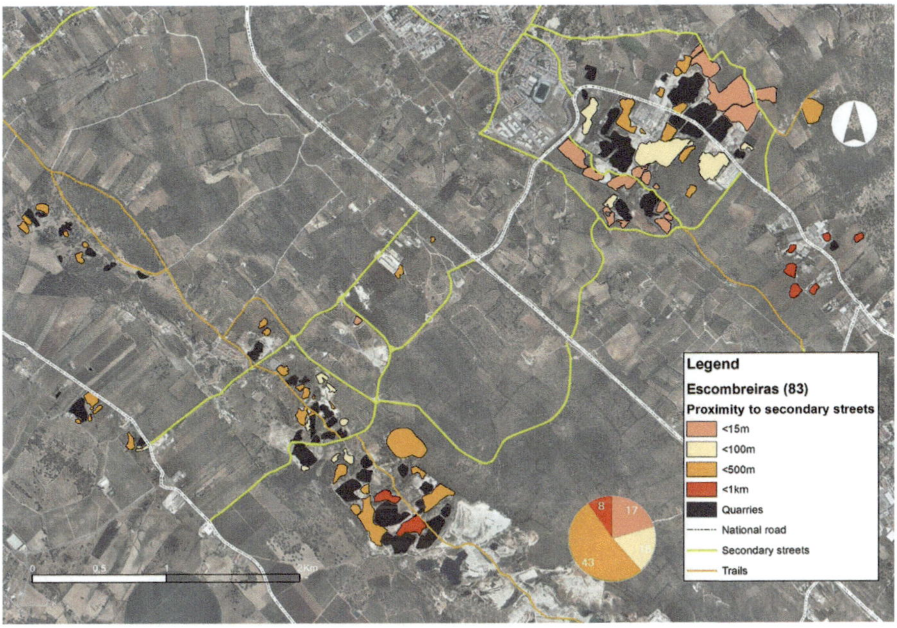

Fig. A22 *Escombreiras*' proximity to secondary streets. (Katia Talento, 2021)

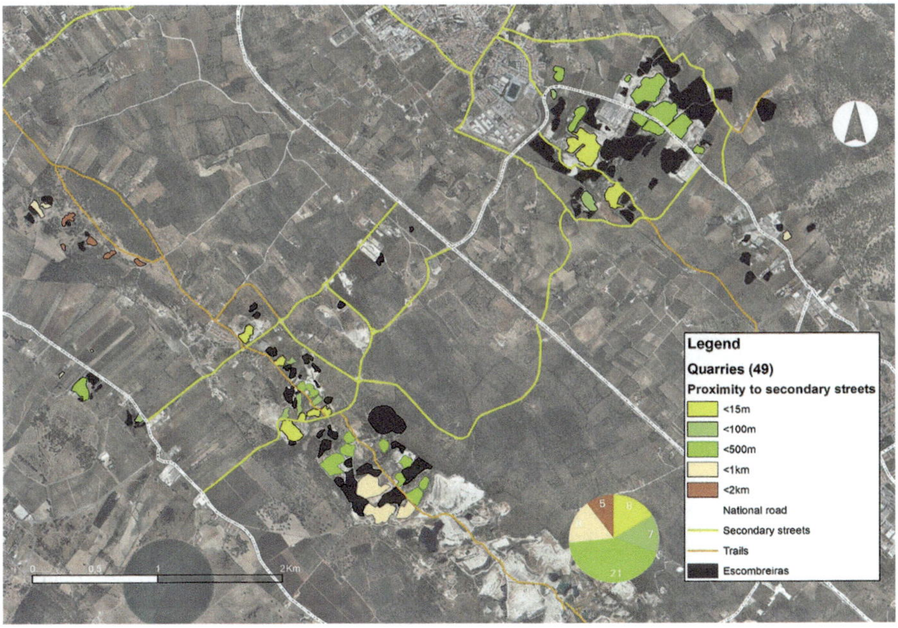

Fig. A23 Quarries' proximity to secondary streets. (Katia Talento, 2021)

Proximity to the Urban Centre (Figs. A24 and A25)

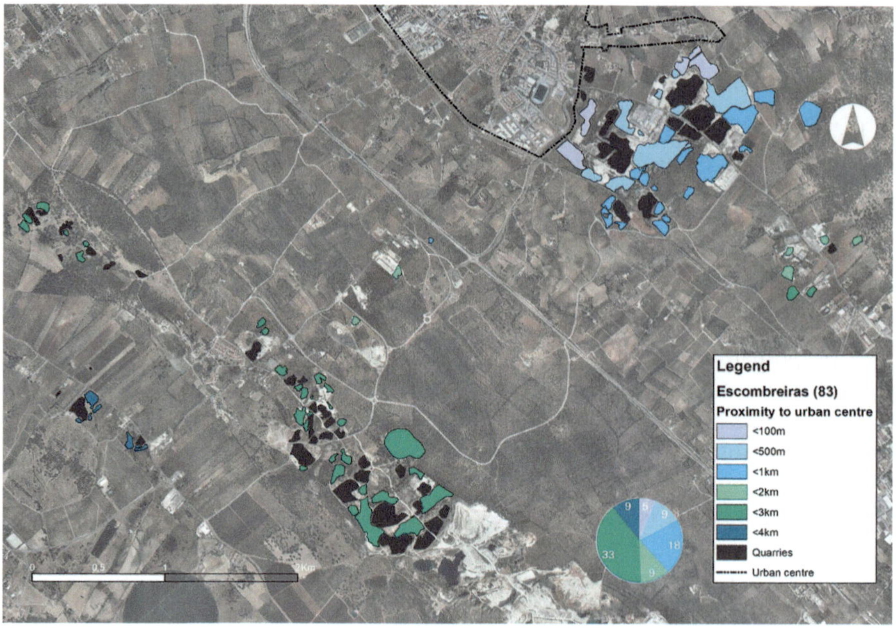

Fig. A24 *Escombreiras*' proximity to the urban centre. (Katia Talento, 2021)

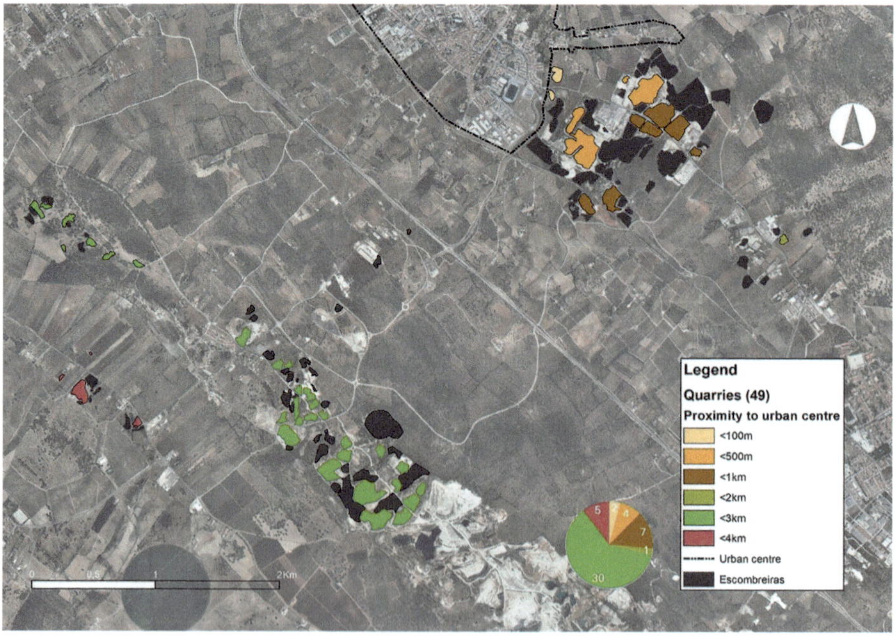

Fig. A25 Quarries' proximity to the urban centre. (Katia Talento, 2021)

Context (Figs. A26 and A27)

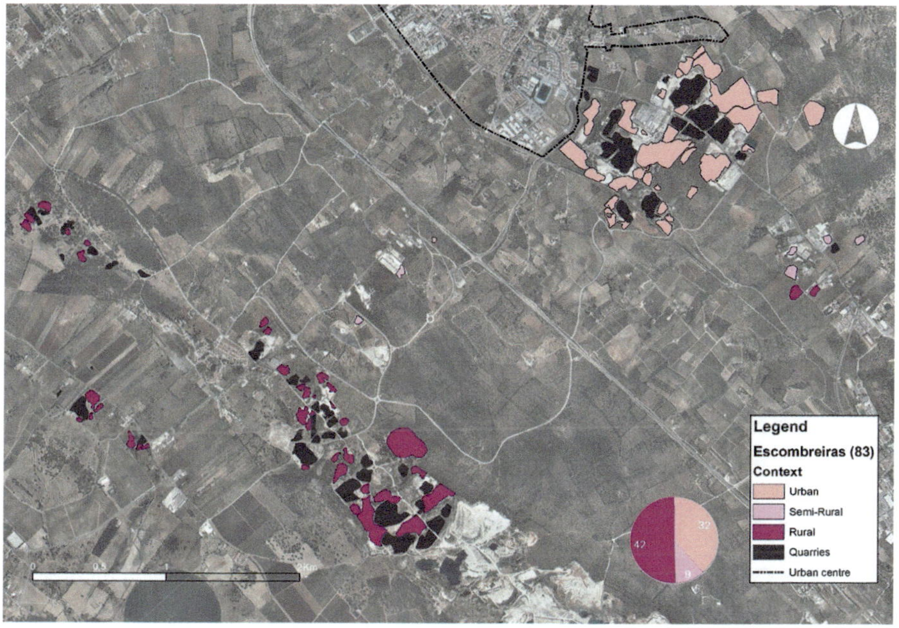

Fig. A26 *Escombreiras*' context. (Katia Talento, 2021)

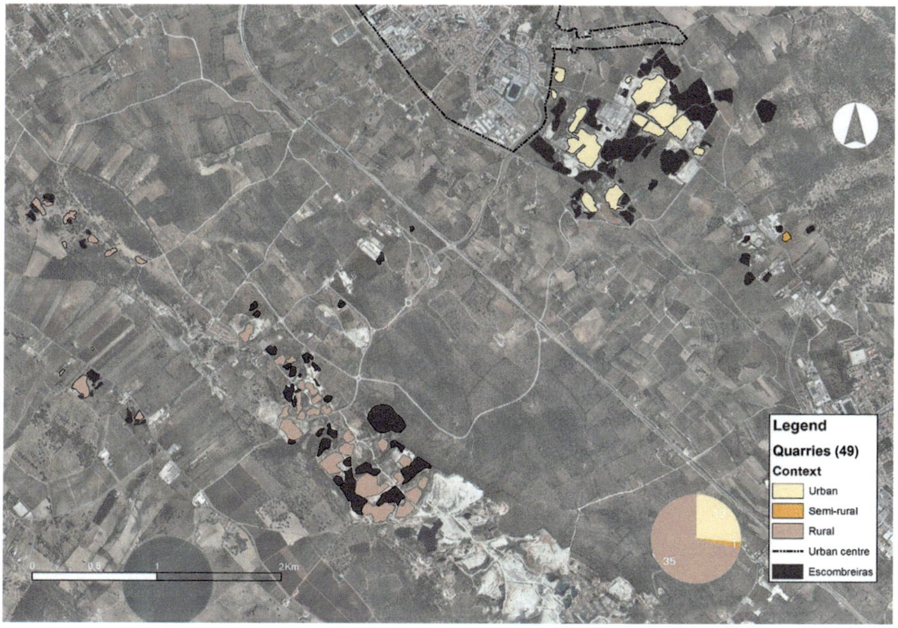

Fig. A27 Quarries' context. (Katia Talento, 2021)

Visual Impact (Fig. A28)

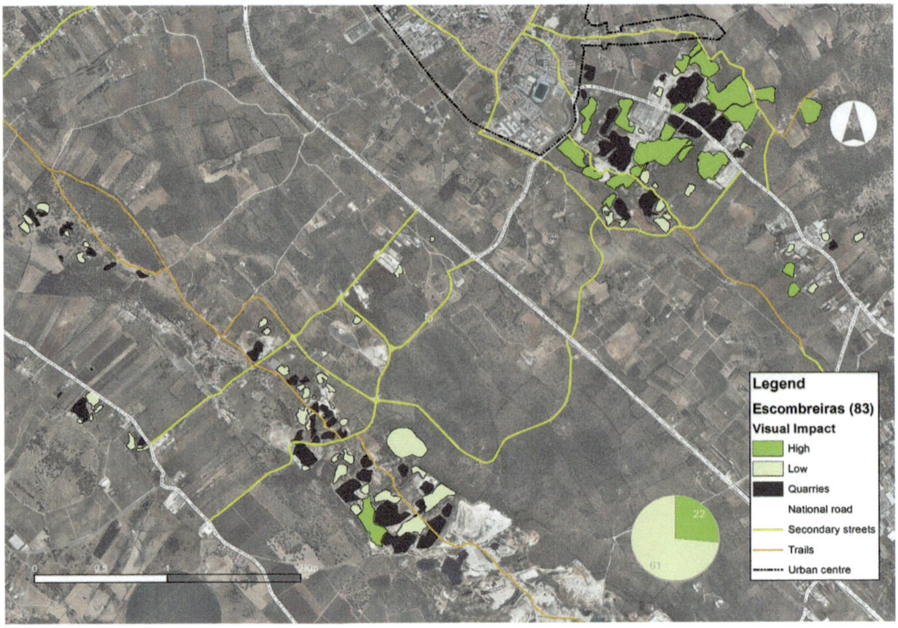

Fig. A28 *Escombreiras*' visual impact. (Katia Talento, 2021)

Main Attractions (Fig. A29)

Fig. A29 Main attractions of Borba. (Katia Talento, 2021)

Risk Condition (Fig. A30)

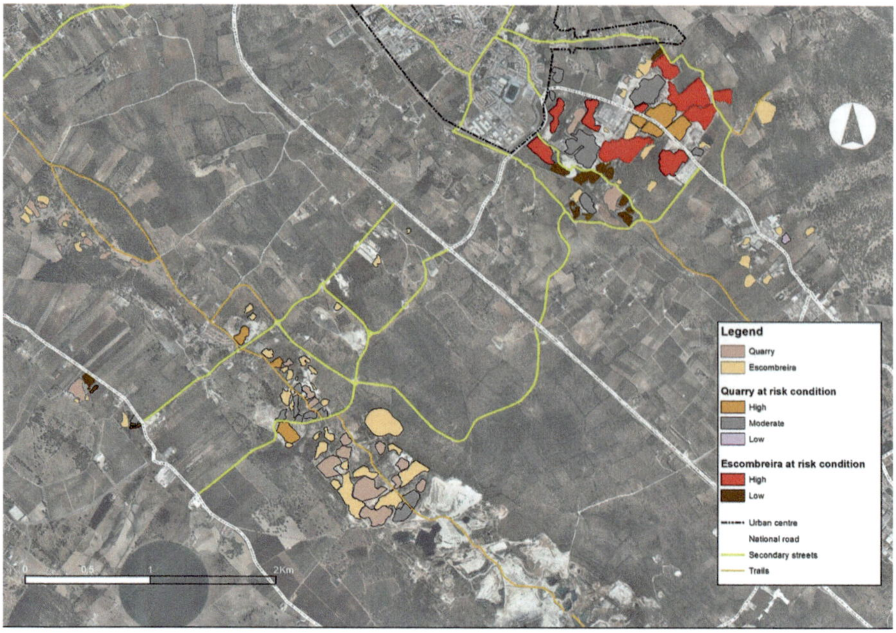

Fig. A30 *Escombreiras* and quarries subjected to a risk condition. (Katia Talento, 2021)

Index

K. Talento, *Sustainable Buildings from Marble Waste Heaps*,
https://doi.org/10.1007/978-3-031-72146-5

GPSR Compliance

The European Union's (EU) General Product Safety Regulation (GPSR) is a set of rules that requires consumer products to be safe and our obligations to ensure this.

If you have any concerns about our products, you can contact us on ProductSafety@springernature.com

In case Publisher is established outside the EU, the EU authorized representative is:

Springer Nature Customer Service Center GmbH
Europaplatz 3
69115 Heidelberg, Germany

Batch number: 10755227

Printed by Printforce, the Netherlands